ESOPTRICS:
THE LOGIC OF THE MIRROR

**(The Divine Algebraic Logic Used By God
To Create And To Maintain The Universe)**

✝✡✝✡✝✡✝✡✝✡✝✡✝✡✝✡✝✡✝✡✝✡✝✡✝✡✝✡✝✡

ESOPTRICS:
THE LOGIC OF THE MIRROR

(The Divine Algebraic Logic Used By God To Create And To Maintain The Universe)

By

Edward N. Haas

AuthorHouse™
1663 Liberty Drive
Bloomington, IN 47403
www.authorhouse.com
Phone: 1-800-839-8640

First published by AuthorHouse 11/19/2009

ISBN: 978-1-4490-4856-3 (e)
ISBN: 978-1-4490-4855-6 (sc)

Printed in the United States of America
Bloomington, Indiana

This book is printed on acid-free paper.

Library of Congress Control Number: 2009912498

ABOUT THE BOOK

Many former thinkers insisted Earth is flat. Even more of them insisted Earth is the center around which the Sun, planets and stars revolve. In each case, they all insisted they were describing what they *observed* with their senses.

We now know Earth is round and, together with the planets, orbits the Sun. The former thinkers, then, could not have *observed* either a flat Earth or one circled by the Sun, planets, and stars. Their descriptions were but *inferences* they failed to tell from *observations*.

These days, virtually every thinker will insist they *observe*: (1) that space and its occupants are infinitely divisible because physically extended, and (2) that time and locomotion are infinitely divisible because absolutely continuous. Esoptrics charges such thinkers, too, are merely promoting *inferences* they fail to tell from *observations*.

To support its allegation, Esoptrics—on one hand—distinguishes between 2 kinds of indivisibles (*i.e.:* points). Indivisibles of the first kind are durationless and indivisible in both space and principle; those of the second kind are indivisible in space but divisible in principle and have a duration which is intermittently outside of time as we experience time. In the latter, Esoptrics does what no one else in history has ever come close to doing—namely: explain how there is a kind of point which—because an astronomical number of principles are conjoined in it—is *logically* divided in principle into an astronomical number of points each *logically* separated in principle from all the others. It's merely a case of understanding that the universe's ultimate constituents are so different from the bodies sensation observes, they can be separate from one another by virtue of nothing more than the internality of what Esoptrics calls their *primary* acts—a concept with no equivalent or precedent in Science. Each primary act is a definite quantity and combination of 6 modes of power (*i.e.:* 3 pos. & 3 neg.) each of which admits of 2^{256} (*i.e.:* 1.1579 x 10^{77}) different intensities.

In further support of its allegation, Esoptrics presents a very unique cosmological model which, strangely enough, matches current models on enough points to make one wonder if it might somehow be correct. The model is fairly elaborate and rather well described in the terms of simple Algebra. Above all, there is, in the model, no trace of either: (1) *physically* extended space and bodies, or (2) infinite space, or (3) infinitely divisible time, space, matter or locomotion.

So far, Esoptrics neither can, nor seeks to, make it *certain* physical extension and infinite divisibility are <u>not</u> a part of our universe. It merely seeks to give enough evidence to make every thinking individual admit they can no longer be *certain* that physical extension and infinite divisibility <u>are</u> a part of our universe.

BOOKS IN PRINT BY EDWARD N. HAAS

Available at any bookstore by special order or on the Internet at web sites such as authorhouse.com or bn.com or amazon.com or others.

Dedicated Primarily To My Brother Gordon
Whose Valuable Input Has Perhaps Shown Me
What Line Of Thought Can Best Serve
As The Spear Head Of
My Effort To Convey Esoptrics To The World.
Dedicated Secondarily To My Other Seven Siblings
And The Parents Who Deigned To Provide Me
With Such Indescribably Helpful Assistants.

In all I shall say in this book, I submit to what is taught by Our Mother, the Holy Roman Church. If there be anything in it contrary to that, it shall be without my knowledge. Therefore, for the love of Our Lord, I implore the learned men who are to examine this book to look at it very carefully and to amend any faults of that kind which may be herein and the many others which it shall have of other kinds. If there be anything good in it, let that be to the glory and honor of God and in service to His most sacred Mother, our Patroness and Lady, whose habit—though all unworthily—I wear.

——**ST. TERESA OF AVILA** (1515 – 1582 A. D.): ***Book Called Way Of Perfection*** as given in the "Protestation". My re-translation – ENH.

CONTENTS

CONTENTS

CONTENTS

CONTENTS

Document #7:

CURVILINEAR MOTION. ..... 147.

CONTENTS

CONTENTS

CHARTS

CONTENTS

CONTENTS

FORMULAS

POSTULATES

✝✡✝✡✝✡✝✡✝✡✝✡✝✡✝✡✝✡✝✡✝✡✝✡✝

LEGAL NOTICE OF PERMISSION TO QUOTE IN A SCHOLARLY DISSERTATION

In particle physics we may have to accept an arbitrary, complicated, not very orderly set of facts, without seeing behind them the harmony in terms of which they might be understood. It is the special faith and dedication of our profession that we will not lightly concede such a defeat.

——**J. ROBERT OPPENHEIMER:** *The Mystery Of Matter* as found on page 68 of *Adventures Of The Mind* from *The Saturday Evening Post* and published by Alfred A. Knopf, New York, 1960. © 1959 by The Curtis Publishing Company. [The scientist may throw in the towel, but the mystical philosopher *__NEVER__*!]

PREFACE:

I:

Human history, says this book, presents us with five colossal cosmological blunders. Two of them have long since bitten the dust. Three are still going strong.

The first blunder to bite the dust was the notion that Earth is flat. Among the masses, this notion may have thrived up until around the time of Christopher Columbus. Among the more sophisticated of thinkers, it died perhaps 400 years B.C.. For example, in his work **On The Heavens**, Aristotle (384-322 B.C.) states that Earth is "necessarily spherical". The advocates of a flat Earth supported their conviction on the grounds that it was an incontrovertible fact they could observe and confirm with their senses.

The second blunder to bite the dust was the notion that Earth is the motionless center of the universe around which all the other heavenly bodies rotate. Up until around the start of the 17th century A.D., virtually everyone—including the vast majority of the most sophisticated thinkers (Aristotle too)—looked upon this geocentric view of the universe as an incontrovertible fact they could observe and confirm with their senses.

This book says the other three blunders are these: (1) Because time is observably continuous, it is infinitely divisible; (2) because every locomotion—such as that of the Earth around the Sun—is observably continuous, every such locomotion is infinitely divisible; and (3) because space and its occupants are all observed to be physically extended, they are infinitely divisible.

This book says that what's confirmed by observation according to these last 3 blunders is no more confirmed by *observation* than was the flatness of Earth or its motionlessness at the universe's center. As a result, physical extension and the infinite divisibility of time, space, matter, and locomotion are no more incontrovertible, scientific, empirical facts than were the flatness of Earth or its motionlessness at the universe's center. Sadly, despite knowing how wrong the supporters of the first 2 blunders were about what's an incontrovertible, scientific fact confirmed by observation, the vast majority of even the most sophisticated of philosophers and scientists—*past or present!*—were and are unshakably sure that physical extension and the infinite divisibility of time, space, etc. are incontrovertible, scientific, empirical facts confirmed by observation.

No, I am not the first or only one to attack the idea of infinitely divisible time, space, matter, and locomotion. There were at least Zeno of Elea (c.495-c.430 B.C.) and Gottfried Wilhelm Leibniz (1646-1716 A.D.). They all, though, did nothing but point out the mind-boggling paradoxes inherent to the very idea of infinite divisibility. Never were they able to present a viable explanation of how time, space, matter, and locomotion could be only finitely and not infinitely divisible, and that was because they were never able to present even the slightest trace of an explanation of how the universe's various locations and material parts could *appear* to be physically outside of one another in space without *actually* being so. With no idea of what *logical* extension is or how it could replace *physical* extension, their arguments against infinite divisibility could do little more than fall on indifferent eyes and ears. What few they did prompt to reply merely puzzled: "A divisibility which is *finite only* is impossible, unless the universe is composed of points; but, whatever their number, points—because they are devoid of length, width, and depth—can't add up to anything except a single point devoid of length, width, and depth. How, then is divisibility *finite only*, unless the entire universe is a single point with no trace of length, width, or depth? But, if the universe is a system comprised of multiple, simultaneous events, one can hardly see how such a manifold system could be embodied in a single point devoid of length, width, and depth."

Oh, how different it is when we come to Esoptrics! For the first time in history, Esoptrics gives the world a highly detailed and mathematically precise explanation of: (1) how apparently continuous time and locomotion can be discontinuous, (2) how the universe's various locations and material parts can be *logically* outside of one another *in the order of sequence* rather than *physically* outside of one another *in space*, (3) how there can be 2 kinds of time—one of which never varies the rate at which it pulses; the other of which can vary that rate astronomically, and (4) how there can be 2 radically different kinds of indivisibles—one of which is durationless and without length, width, or depth and, therefore, is absolutely indivisible whether in space or in principle or in any other way whatsoever; the other of which is indivisible in space but divisible in principle and has a kind of duration which is intermittently outside of time as we experience time. Grasp the basic principles of Esoptrics, and one cannot fail to see exactly how the universe's manifold system can indeed be embodied in a point logically divisible into an astronomical number of points each logically separate from all the others.

Be fair about it, and one must admit that the explanations—which Esoptrics brings to bear against physical extension and infinite divisibility—are one of the greatest achievements by far in the history of Cosmology and in the whole history of humanity's intellectual advancement. For, prior to Esoptrics, there was never the slightest trace of an explanation of either: (1) how apparently continuous time and locomotion could be discontinuous, or: (2) what a *logically* extended point is, or: (3) how such a point could contain the universe's manifold system, or: (4) how there could be two radically different kinds of indivisibles or: (5) how there could be 2 radically different kinds of time—a variable versus an invariable kind. Esoptrics, though, explains it all in great detail and with mathematical precision.

Some will say that a noteworthy cosmological theory must give some proof of its validity, and, to do that, it must make accurate predictions regarding what future observations shall observe. I gladly accept that as true with regard to any theory seeking to prove its cosmological model correct. I deny it's true with regard to a theory seeking only to prove that physical extension and infinite divisibility are not either empirical, scientific facts confirmed by observation or inferences too sacred to be doubted. To achieve those 2 goals, Esoptrics need meet only 2 requirements:

(1) produce a cosmological model which—while apparently possible rather than patently impossible—agrees on a large number of points with what today's Science says about the universe, and (2) is a cosmological model in which physical extension and infinite divisibility have no part. For, once Esoptrics produces such a model, it produces evidence strong enough to require every thinking individual to admit: "We can no longer be certain our universe is one in which physical extension and infinite divisibility have a part."

To confirm Esoptrics has met the first requirement, one need only read and understand Esoptrics' descriptions of the various components of its cosmological model. To confirm Esoptrics has met the second requirement, one must first read and understand Esoptrics' description of a *logically* extended point. That means reading and understanding what Esoptrics says about primary acts (*i.e.:* a concept without precedent in Science) and: (1) "physically outside of one another in space", vs. (2) "logically outside of one another in the order of sequence" (*i.e.:* yet another concept without precedent in Science). Once that is grasped, it will be "easy as pie" to understand how apparently continuous time and locomotion can be discontinuous. It will also be easy to understand how God can be omnipresent.

Once that is done, it makes no difference either: (1) how far off the mark Esoptrics is as a completely accurate cosmological model, or: (2) how devoid of proof and predictions it might be. It shall still have achieved its main goal, and must leave every fair minded philosopher and scientist willing to admit: "In the face of Esoptrics, we can no longer be certain that physical extension and infinite divisibility are ***not*** just 3 more colossal cosmological blunders."

II:

This book is not exactly a single, coherent work. Much of it is composed of several *independent* attempts to set forth Esoptrics' basic concepts. As a result, the reader will find there is much repetition here.

That, though, may not be a bad idea. That's because of the extent to which *physical* extension and *infinite* divisibility are ingrained in the human psyche. Human beings are so used to those 2 principles, they automatically and instinctively reject *logical* extension and *finite only* divisibility with great intensity. So great is that intensity, it's not even remotely possible for them to follow any explanation of the latter 2 principles, unless more than one way of phrasing the explanation is thrown at them over and over again.

AT HAASWOOD, LOUISIANA, THURSDAY, OCTOBER 1, 2009.
AD MAJOREM DEI GLORIAM.

His In Omnibus Rebus
Judicium Praevaleat
Ecclesiae Catholicae
Quod Est,
Judicio Ecclesiae Catholicae,
Judicium Cathedrae Petri.

Document #1:

ESOPTRICS[1] VS. 3 SACRED COWS:

PART I: SOME OF ESOPTRICS' BASIC ASSERTIONS:

A. A COMPARISON WITH THE GEOCENTRIC UNIVERSE:

In the thousands of years from the dawn of recorded history to the mid six-teenth century—after a Catholic cleric named Nicolaus Copernicus (a/k/a Mikolaj Kopernik 1473-1543) turned enough heads in the opposite direction—the vast ma-jority of even the most educated and brilliant people in the world had not the slightest doubt that Earth is the motionless center of the universe around which all the other heavenly bodies rotate. For all but perhaps a handful, such a *geocentric* view of the universe was a sacrosanct, utterly unassailable, foregone conclusion (***i.e.:*** a sacred cow). Had you asked them why, they—one way or another—would at least effectively have said: "It's an empirical, scientific fact anyone can confirm for one's self by merely turning one's gaze to the heavens and then carefully observing what is present to one's eyes." [2]

---

[1] "Esoptrics: The Logic Of The Mirror" is the name I give to the cosmological theory I've been work-ing on since perhaps as early as 1957. "Esoptrics" comes from the Greek word for "mirror".

[2] In Chapters 13 and 14 of Book II of his work **On The Heavens**, Aristotle (384-322 BC) explores the issues of Earth's shape, motion, and location. An English version is given on pages 384-389 of vol. 8 of **Great Books Of The Western World**. On page 388, he repeatedly assures us Earth is at "the cen-ter of the whole", moves not, and is necessarily spherical. However, on page 384 near the beginning of chapter 13, he tells us that—though Earth is regarded as at the center by all who say the whole heaven is finite—the Pythagoreans maintained fire is at the universe's center and Earth one of the "stars" orbiting that fiery center. Some add that this fiery center was not the Sun in Pythagorean thinking, since, for them, the Sun, too, was one of the bodies orbiting the central fire. Since we now know the Sun orbits our galaxy's fiery center, the Pythagoreans, it seems, were an astonishingly pre-scient group. On the said page 384, Aristotle dismisses the Pythagoreans as trying to force observed facts to comply with their theories rather than looking for theories which account for observed facts (It's often said the Pythagoreans worshiped fire and thus, for theological reasons, had to see it as the center of everything.). How ironical that those—those trying to force observed facts to comply with what some would call their "theological superstitions"—got it *somewhat* right 2,500 years ago; whereas, their opposites—namely: those who made much ado about sticking to the observed facts—didn't get it right *at all* until a little over 2,000 years later! Some sources say Aristarchus of Samos (c. 270 BC) and Seleucus of Seleucia (c.150 BC) also supported Heliocentrism against Geocentrism.

1

We now know they were mistaken. Their geocentric universe was ever nothing but a fantasy. It was ever an erroneous *inference* they mistakenly saw as an *observable fact* clearly present in the images their senses presented to them. As such, it was a purely subjective fabrication present only in the minds of people: (1) who failed to distinguish between what they *actually observed* and what was merely an *inference* they themselves were reading into, and thus *causing*, the observed to imply, and (2) who, as a result, jumped too quickly from the intra-mental evidences in their senses to the extra-mental cause of the images in their senses.

How does #1 entail #2? An observed fact is *ipso facto* certain and, as such, needs no corroboration. Oppositely, what is merely an inference *implied* by an observed fact requires corroboration by even more observed facts all implying the same inference. Indeed, if there's any evidence *against* the proffered inference, the need for corroboration can mount dramatically depending upon the gravity of the contradicting evidence.[1] Without such corroboration, one can't be certain the inference is not merely an implication of one's own invention rather than an inference actually implied by the observed fact itself. If one fails to see one is dealing with an inference rather than an observed fact, one will fail to see the need for corroboration and will "jump the gun", as they say. It's the inevitable result of assuming one is dealing with observable fact rather than with an inference one is reading into the observed and thus *causing* it to imply. The Geocentric Theory—like the notion of a flat Earth—is a perfect example of that. [2]

Esoptrics now says there are 3 other supposedly sacrosanct, utterly unassailable, foregone conclusions (*i.e.:* sacred cows) which—like the geocentric universe and a flat Earth—are purely subjective fabrications present only in the minds of people who fail to notice that what they deem observable facts are actually nothing more than erroneous inferences they themselves are forcing the observed facts to imply. Those 3 sacred cows are: (1) physically extended space and spatial objects, (2) continuous time, and (3) continuous locomotion.

B. REGARDING THE FIRST SACRED COW, PHYSICAL EXTENSION:

B-1: PRELIMINARY REMARKS ABOUT FORMS & GENERATORS:

In the hope of making it at least somewhat less difficult to keep up with what is to follow, let me start with these preliminary remarks. With these, I will seek to give my reader simple drawings *symbolically* explaining what Esoptrics means by: (1) *logically* extended space and spatial objects, and (2) generators and forms.

---

[1] Give or take a year, I was 20 when I first read a description of what Zeno of Elea (c495-c430 BC) had to say about infinite divisibility. From that day to this, I remain firmly convinced that, 2,500 years ago, Zeno presented evidence so gravely contradicting the notion of infinite divisibility, as to make the notion utterly untenable. Yes, I've read replies to Zeno's paradoxes; but, they merely left me marveling how such authors could fail to see in a flash that infinite divisibility is, on its face, self-contradictory gibberish as manifestly such as is a square circle.

[2] Flat Earth advocates used basically the same tactic to justify their conviction—namely: They appealed to what they could observe with their senses. Science's "empirical" evidences often suck.

Esoptrics says every ultimate constituent of the universe is a single *micro*-scopic one and a single *macro*scopic one forever joined *logically* at their centers. It calls every *micro*scopic ultimate a "carrying generator" or "generator" for short. It calls every *macro*scopic ultimate a "piggyback form" or "form" for short. Every generator is a *carrying* generator, because it's actuality is forever logically centered upon its piggyback form's actuality. Every form is a *piggyback* form, because its actuality is forever logically centered upon its carrying generator's actuality. Every generator is called such because its every act generates a unit of force in the sense of tension and vibration—something forms do not do. Remember that well.

Esoptrics says no ultimate constituent has a *size* in the final analysis. As some might prefer to hear it expressed: The "size" of any given ultimate is merely an apparent (*i.e.:* phenomenal) rather than a *real* (*i.e.:* noumenal) characteristic. To put it another way: *For our human senses*, the universe's ultimate constituents have different sizes, because that's the only way our *senses* can experience them. In reality, though (says Esoptrics), the universe's ultimates have different levels of power[1] rather than different sizes. As a result, what our senses experience as an instance of "small in size" is actually one of "lower in power", and, what our senses experience as an instance of "large in size" is actually one of "higher in power".

Every ultimate, at any given instant, is performing a kind of act so unusual, it exhibits no kind of change or motion known to us—particularly not any kind of locomotion. Instead, each such act is a definite level of power as long as it's the same, one act. Hereafter: **"ACT" ALWAYS = "PRIMARY ACT"**, which is to say this kind of unusual act. Be doubly, doubly sure to keep that in mind. It's super critical.

Esoptrics says there are 2^{256} (*i.e.:* 1.15792×10^{77}) levels of power. Esoptrics names those levels OD1, OD2, OD3, etc., thru and including $OD2^{256}$. "OD" stands for ontological distance, which is the number of steps from zero power (*i.e.:* zero Being). At OD1, every act's power is one step above zero. At OD2, every act's power is 2 x that of every act at OD1. At OD3, every act's power is 3 x that of every act at OD1, etc., until, at $OD2^{256}$, every act's power is 2^{256} x that of every act at OD1.

Let the letter Z represent the level of power at $OD2^{256}$. In that case, the level of power at OD1 is $2^{-256}Z$ (Note the minus sign.); the level at OD2 = $2(2^{-256}Z)$; the level at OD3 = $3(2^{-256}Z)$; the level of power at OD4 = $4(2^{-256}Z)$; and so forth.

Every act of every *generator* always has a power level of OD1. Only among the *forms*, do we find acts in which the level of power is either OD1 or OD2 or OD3 or OD4, etc., to and including $OD2^{256}$.

As I said, for Esoptrics, ultimates do not *actually* have size; but, to communicate its principles in a convenient manner, I present a series of pictures so: To simplify the notion of a form performing an act at OD1, imagine a cube whose length, breadth, and depth (LB&D for short) are each roughly 10^{-47} cm. (*i.e.:* 2^{-129} x the dia. of the hydrogen atom with its electron in its outermost orbit); to simplify the notion of a form performing an act at OD2, imagine a cube roughly $2(10^{-47})$ cm. LB&D; etc., until, at $OD2^{256}$, the cube's LB&D are each roughly 10^{31} cm., which is to say roughly 18 trillion light years (*i.e.:* 1.798×10^{13} l. y. to be more precise). For Esoptrics, $7.3468398 \times 10^{-47}$ cm. = 1 alphatopon or 1Φ for short. Nothing is smaller.

---

[1] I'd say different "intensities of Being"; but, these days, the word "Being" is too much of a pariah.

ESOPTRICS: LOGIC OF THE MIRROR

A cube???!!! Why would one imagine a cube? First, Esoptrics says there are *three* kinds of power. Being the logic of the *mirror*, Esoptrics deals with reverse images and, so, because it says there are 3 kinds of *positive* power, it must say there are three kinds of *negative* power. In other words: Each of the 3 positive powers has a reverse (*i.e.:* mirrored) power. As a result, when we speak of the power available to generators and forms, we're actually talking about 3 power/anti-power combos available to them. That gives us a total of 6 "modes" of power, if I may so speak.

Esoptrics names the 3 positive powers: (1) A, (2) A' (*i.e.:* A prime), and (3) B' (*i.e.:* B prime). Esoptrics names the 3 negative powers: (1) B, (2) C, and (3) C' (*i.e.:* C prime).[1] The negative power called B is the anti-power to the positive power called A and vice versa. The negative power called C is the anti-power to the positive power called A' and vice versa. The negative power called C' is the anti-power to the positive power called B' and vice versa. As a result, the 3 power/anti-power combos are: (1) A vs. B, (2) A' vs. C, and (3) B' vs. C'.

For the sake of convenient communication and sensory perception, we thus have 3 *axes*—namely: (1) AB, (2) A'C, and (3) B'C'. Add in that each of these axes is logically related to the other two in the same manner, and the best way to *picture* that relationship is to imagine each axis perpendicular to the other two. If you do that, you are left with the picture of a cube.

Whence come these 6 modes of power? Esoptrics says it depends upon whether you're talking about forms or generators.

FOR EVERY FORM, they come from the 6 points of reference found in an infinite level of power which, though technically outside the universe, "calls" the forms to its infinite level by, so to speak, giving them a finite foretaste of its 6 modes of infinite power. As Esoptrics prefers to express it: Every form is *in potency* to infinity 6 ways and thus has 6 modes of power.

FOR EVERY GENERATOR, its 6 modes of power come from some form which is other than the generator's piggyback form. As Esoptrics prefers to express it: For any given generator, its 6 modes of power come from the form to which the given generator is *in potency*.

For every generator, then, 2 forms are important: (1) its piggyback form, and (2) the form to which it is in potency. The OD of the former determines what OD its carrying generator shall seek in the latter. The latter—in providing a generator with its 6 modes of power—also provides it with a field of locomotion. Every form has only 1 carrying generator, but may have many generators in potency to it.

Every *form*, in each of its primary acts, actualizes—and equally so—a definite fraction of each of the 6 modes available to it. In each of its primary acts, no *generator* is ever able to actualize—in the same one act—both a power and its anti-power. In each of its acts, then, no generator is ever able to actualize—in the same one act—more than 3 of the 6 modes available to it, and it may well actualize only 1 in the same one act, or it may well actualize only 2 in the same one act, as long as those 2 modes are not opposites. Let's use pictures to explain *symbolically* this all-

---

[1] I shouldn't have to spell out, here and now, why I chose such names; and so, I will not do so until much later in a document I've entitled: "Esoptrics' Foundation In Purely Introspective Reasoning". Too late now, I wish I had named the positive powers A, B, & C and the negatives A', B' & C'.

important difference. To that end, let's turn to drawings #1, 2, 3, and 4.

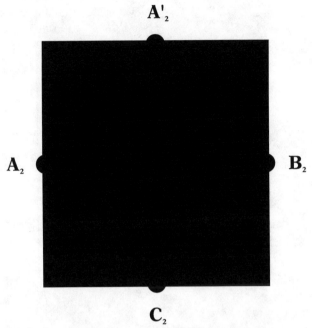

DRAWING #1: A FORM PERFORMING AN ACT AT OD2.
ITS ACT'S LEVEL OF POWER IS THUS:
$$[2(2^{-256}Z)A] + [2(2^{-256}Z)B] + [2(2^{-256}Z)A'] + [2(2^{-256}Z)C] + [2(2^{-256}Z)B'] + [2(2^{-256}Z)C']$$

As you can perhaps readily see, drawing #1 is of a square; and yet, it needs to be the drawing of a cube. Since I do not have the skill required for one to draw a cube on a flat surface, we'll have to be content with a square which we then convert into a cube by imagination. To achieve that conversion, first imagine the drawing is sitting flat on the top of a table. Next, imagine the black area expands upward into the air above the table toward a reference point called B', and the distance from the table to the top of the upwardly expanded black area equals the distance from the center of the square to the center of the dot marked B_2. Finally, imagine the black area expands downward into the air below the table toward a reference point called C', and the distance from the table to the bottom of the downwardly expanded black area equals the distance from the center of the square to the center of the dot marked B_2.

Drawing #1 is by no means a picture of what a form "looks like". Wholly to the contrary, it is merely a kind of allegory serving to explain *symbolically* what it means to say the form is performing an act at power level OD2 with regard to each and every one of the 6 modes of power. In other words: In this particular one of its acts, the form's level of power is $2(2^{-256}Z)$ with regard to A, A', B, B', C, and C'.

Drawing #2 on the next page should, I hope, make extremely evident the contrast between the act of a form and the act of a generator. Drawing #2 tells us nothing about the generator's piggyback form; and so, the white square represents the act of the form to which the generator is in potency. The white square is thus the same as the black area of drawing #1.

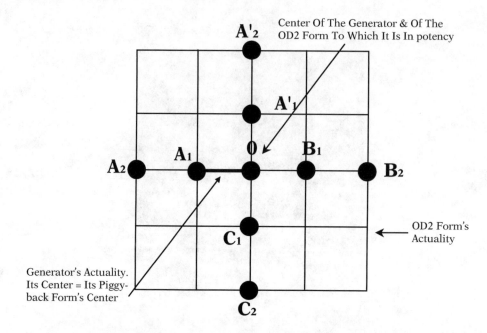

Center Of The Generator & Of The OD2 Form To Which It Is In potency

OD2 Form's Actuality

Generator's Actuality. Its Center = Its Piggy-back Form's Center

DRAWING #2: A GENERATOR PERFORMING A "ONE DIMENSIONAL" ACT AT OD1 BUT WITHIN THE CONFINES OF THE ACT BEING PERFORMED BY A FORM AT OD2. ITS LEVEL OF POWER THUS = $(2^{-256}Z)A$

Because the generator is performing an act at OD1, its act's level of power is $2^{-256}Z$. Because its act is "one dimensional", its unit of force is too, and its act's level of power is $2^{-256}Z$ only with regard to the positive power called A. We express that *symbolically* by making the line $A_1 0$ thicker than the other lines.

Were we to use drawing #2 to depict symbolically the generator's *piggyback* form (as opposed to the form to which it's in potency), we'd first have to picture a black area whose length, breadth, and depth are each one-half that of the black area in drawing #1. Next, we'd have to place the center of that black area at the center of the line $A_1 0$ (See drawing #7 on pg. 32.). Yes, that means the generator's piggyback form is performing an act at OD1; and so, the power level of its act is $2^{-256}Z$ with regard to each of the modes A', A', B, B', C, and C'. A piggyback form at OD1 always causes its carrying generator to be in potency to a form at OD2.

As in drawing #2, drawing #3 (next page) tells us nothing about the generator's piggyback form; and so, the large white square is a cube to imagination and represents the act of the form to which the generator is in potency. The large white square/cube is thus the same as the black area of drawing #1.

The small black square in drawing #3 is not a cube and symbolically expresses a *second* act on the part of the same generator we symbolically described in drawing #2. It's just that, on this go around, the act and its unit of force are *two*-dimensional. As a result: (1) Because the generator is performing an act at OD1, its

6

act's level of power is 2^{-256}Z; but: (2) because its act is "two dimensional", its act's level of power is 2^{-256} Z with regard to both the positive power called A and the positive power called A'. That latter result is, of course, the reason why the drawing presents us with a black *square* rather than a black *line*.

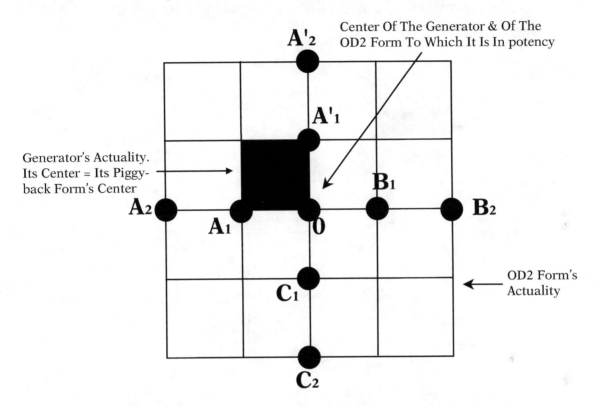

DRAWING #3: A GENERATOR PERFORMING A "TWO DIMENSIONAL" ACT AT OD1 BUT WITHIN THE CONFINES OF THE ACT BEING PERFORMED BY A FORM AT OD2. ITS LEVEL OF POWER THUS = $(2^{-256}Z)A + (2^{-256}Z)A'$

Were we to use drawing #3 to depict symbolically the generator's *piggyback* form (as opposed to the form to which it's in potency), we'd first have to picture a black area whose length, breadth, and depth are each one-half that of the black area in drawing #1. Next, we'd have to place the center of that black area at the center of the small black square in drawing #3.

As in drawings #2 & #3, drawing #4 (next page) tells us nothing about the generator's piggyback form; and so, the large white square/cube represents the act of the form to which the generator is in potency. The large white square is thus the same as the black area of drawing #1, and the dot marked "Generator's Center" is also the center of the black area of drawing #1.

The small black *square* in drawing #4, not a *cube*, symbolically expresses a *third* act of the same generator we symbolically described in drawings #2 & #3. As

7

in drawing #3, the act and its unit of force are *two*-dimensional; and so: (1) Because the generator is performing an act at OD1, its act's level of power is $2^{-256}Z$; but: (2) because its act is "two dimensional", its act's level of power is $2^{-256}Z$ with regard to both the positive power called A and the positive power called A'. That latter result is why the drawing presents us with a black *square* rather than a black *line*.

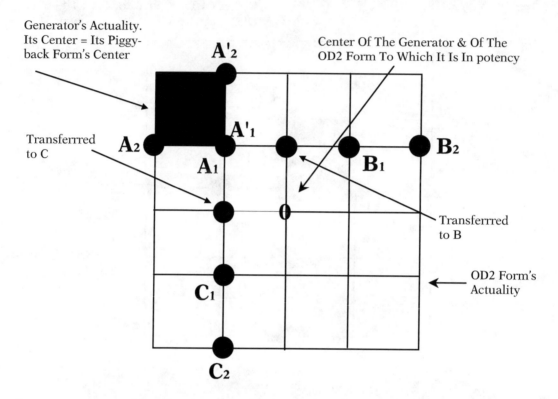

DRAWING #4: A GENERATOR PERFORMING A "TWO DIMENSIONAL" ACT AT OD1 BUT WITHIN THE CONFINES OF THE ACT BEING PERFORMED BY A FORM AT OD2. ITS LEVEL OF POWER THUS = $[2(2^{-256}Z)A – (2^{-256}Z)B] + [2(2^{-256Z})A' – (2^{-256}Z)C]$

But, wait a minute! How is the black square now in the upper left hand corner of the larger square rather than exactly where it is in drawing #3? The positive power A and its negative anti-power B are interchangeable. By the same token, the positive power A' and its negative anti-power C are interchangeable. The same can be said of the positive power B' and its negative anti-power C'. The interchange takes place whenever required—and to the extent required—to preserve the power level standard to the OD involved. Therefore, the smaller black square in drawing #4 is in the upper left-hand corner of the larger white square for these two reasons: (1) The generator's level of power is $2(2^{-256}Z)A$; whereas, at OD1, it's required to be $(2^{-256}Z)A$; and so, $(2^{-256}Z)A$ has been transferred over to A's anti-power, B; (2) the generator's level of power is $2(2^{-256}Z)A'$ when, at OD1, it should be $(2^{-256}Z)A'$; and so,

$(2^{-256}Z)A'$ has been transferred over to C, which, of course, is the anti-power of A'. That, after all, is why drawing #4 describes the power level of the generator's act as: $[2(2^{-256}Z)A$ *minus* $(2^{-256}Z)B] + [2(2^{-256Z})A'$ *minus* $(2^{-256}Z)C]$.

To use drawing #4 to depict symbolically the generator's piggyback form, we must first picture a black area whose LB&D are each one-half that of the black area in drawing #1. Next, we must place the center of that black area at the center of the small black square in drawing #4. See drawings #7, 16, 17, 18, 19, and 20 on pages 32, 110, 111, 112, 113, and 114 respectively for further clarification.

Notice that, in drawings #2, 3, and 4, the generator is in potency to a form whose OD is greater than the OD of the generator's own *piggyback* form. That's a standard condition. Every carrying generator is always in potency only to a form whose OD is greater than that of its own *piggyback* form. In other words: For its field of locomotion and source of 6 modes of power, every carrying generator turns to a form whose ontological distance is greater than the ontological distance of the generator's own piggyback form.

Notice also that, in drawings #2, 3, and 4, the center of the generator is not at the center of the generator's act; rather, it's at the center of the form to which the generator is in potency. Quite the opposite is true of every form. For all forms at all times everywhere, the center of the form is always at the center of the form's own act in addition to being at the center of the act of the form's carrying generator.[1]

That has 2 important implications: (1) It means that, *by themselves*, forms cannot engage in locomotion. *By themselves*, they can only expand and contract by moving up and down in OD. But, every form rides piggyback on a generator which, by itself, is capable of locomotion within the confines of the form to which it is in potency; and so, by means of its carrying generator, every form is capable of locomotion within the confines of the form to which its carrying generator is in potency. (2) It means a balance which explains why forms produce no units of force.

But, from the standpoint of any given carrying generator, every form—with an OD greater than that of its piggyback form—is potentially a field of locomotion. It should be rather obvious what that implies—namely: Virtually every act of every form is a *moving* field of locomotion as far as some generator is concerned. Every act of every form is thus what we can call a "*moving* space envelope" or a "*moving* time-space stratum" (a term whose significance will be spelled out much later.). Oh, what a dramatic change that is! How so?!

From time immemorial, it has been commonly said that the universe is space together with its occupants. Space was thus seen as some kind of single entity (or nonentity, as at least some, if not the vast majority, imagined it to be) which, if it did not extend outwardly in all directions forever (*i.e.:* if it is not infinite in diameter), has no inward limit to how small the smallest segment of it is, which is to say it's infinitely divisible. Esoptrics, though, says that the universe is an astronomically immense collection of units of space (*i.e.:* "space envelopes" or "time-space strata", if you prefer) each of which is the act of a piggyback form. What's more, each of these "space envelopes"—by means of its carrying generator—may be moving

---

[1] Notice also the similarity between drawings #2 thru 4 and the current cosmological theory about strings which vibrate. And what about forms versus branes?

course, how Esoptrics explains "light speed" (*i.e.:* $1\Phi/2^{128}K$)[1] is very different from how classical physics does; and yet, the 2 explanations, however different, give us the same velocity under standard conditions. What's more, Esoptrics goes beyond Michelson-Morley and explains why light speed is the same regardless of Earth's direction of travel. It has to do with the fact that, since space is the act of a form, space itself moves. See pages 261-275 of my *On Philosophy: 1 Long & 4 Short*.

Esoptrics says that, as a form, whose native OD is a power of 2, accelerates beyond its native OD, it <u>ingests</u> one form for each of the OD's thru which it passes in the course of its ascent. Naturally, if it then decelerates to some extent, it will then <u>expel</u> one form for each of the OD's thru which it passes in the course of its descent. That's how Esoptrics accounts for electro-magnetic radiation.

As I understand Einstein (perhaps rather poorly), matter and mass come to be when an area of the time-space field contracts and pulls downward to form a kind of knot in the time-space field and at the bottom of a kind of valley in the time-space field. Supposedly, the depth of the valley varies as the mass of the matter (*i.e.:* knot) thus formed. That then allows the time-space field to—so to speak—*push* the less massive occupants of the time-space field down into the more massive valley. I recently heard a History Channel presentation in which the commentator contrasted that with Newton's view of gravity according to which (as with Esoptrics) the larger mass *pulls* the smaller mass toward itself.

However, as I—perhaps wrongly—understand the story, Einstein was never able to explain why any area of the time-space field would contract and form the all-important knot and dip. As I've often heard it expressed: Einstein never came up with a "non-singular solution to the theory of the unified field". For Esoptrics, that's no big problem as a result of the monumentally powerful impact of these 3 of its concepts: (1) what constitutes a piggyback form and, thereby, makes it a field of locomotion, which is to say a time-space envelope;[2] (2) what constitutes the native versus the current OD of a piggyback form; and (3) how accelerating a form beyond its native OD causes many forms (*i.e.:* many units of time-space) to so fuse as to produce an area of space which admits of an astronomical number of time-space strata each concentric with all the others. In short, Esoptrics is here able to surpass Einstein because he knew of only 1 whereas Esoptrics knows of 2^{256} kinds of space.

As I—perhaps wrongly—understand the story, Einstein says that, in the vicinity of a material body, the time-space field is <u>curved</u>. Esoptrics says that, in the vicinity of a material body, the time-space field is <u>stratified</u>, which is to say the time-space field is actually a definite number of time-space strata (*i.e.:* forms) logically concentric with one another. Naturally, the number of time-space strata present in the stratified time-space field increases in proportion to the increase in the mass of the material body producing the stratified time-space field. That's because the mass of every material body is determined by how greatly the body's leading form has accelerated beyond its native OD, and, in turn, the number of time-space

---

[1] 1Φ = 1 alphatopon = $7.3468398 \times 10^{-47}$ cm. = the smallest smallness, and $1K$ = 1 alphakronon = 7.20217×10^{-96} sec. = time's smallest unit. That makes the above = 29,979,245,802 cm./sec..

[2] Yes, I'm well aware I've not yet explained how the act of a form is a *time*-space envelope rather than merely a *space* envelope. All in due course!

strata (*i.e.:* forms) included in the stratification equals the number of ontological distances thru which the leading form has passed as it accelerated.

Esoptrics says that, no matter where you are in the universe, the time-space field is stratified to some extent. In other words: No matter where you are, there are at least several logically concentric "boxes" big enough to permeate that area. But, the closer you get to the center of either a planet or a star or a galaxy or a galaxy cluster, etc., the more stratified does the time-space field become, which is to say the more so *four* dimensional the time-space field is. Drawing #5 below explains why that is so. The largest "box" is the cluster's leading form.

DRAWING #5a: THE STRATIFICATION OF SPACE IN THE VICINITY OF A MATERIAL MASS[1]

As you can see in the drawing, the number of "boxes" permeating a given area of space depends upon how far that area of the universe is from the center of which material mass. For example, in the above drawing, there is a distance from the center at which 14 "boxes" are present (making each place in space 14 places), then a distance at which only 13 are present (making each place 13 places), and so forth. Of course, it would seem, from the above, that there is a distance at which only 1 "box" is present; but, to make the picture more accurate, you must imagine several of the above structures overlapping one another to some extent.

With that, I conclude these preliminary remarks. If what's here is not enough to ease the burden of keeping up with what follows, then I doubt any number of words on my part can achieve that goal.

B-2. *PHYSICALLY* EXTENDED SPACE & SPATIAL OBJECTS:

---

[1] In the case of an atom, the stratification involves roughly 2^{128} "boxes" the largest of which is only 5×10^{-8} centimeters. In the case of the most massive occupant of the universe, the stratification involves $2^{256}-1$ "boxes" the largest of which is 18 trillion light years. In Esoptrics, mass, I suspect, is not merely a case of how many "boxes" are concentric. Mass is also a case of how many acts are performed by each of the smaller "boxes" before the largest "box" switches from one primary act to another one. That *suggests* that, for Esoptrics, mass rises and falls thru a definite range.

As for sacred cow #1: In the thinking of the vast majority of even the most educated and brilliant people, different locations and different parts of space's occupants are outside of one another simply because that's the way physically real things _are_ by their very nature. In other words, for all but perhaps this author, different places and pieces of matter are outside of one another _in space_, because, by its very nature, space is a single, homogeneous nothingness so extended as to encompass every possible level whether of bigness or smallness: Infinitely elongated in diameter (I first rejected this as impossible when I was 7.), there is no limit to how small the smallest segment of it is, which is to say it's infinitely extended outwardly and infinitely divisible inwardly. What an amazing nothingness![1]

Are these two properties of space ones they can confirm by observation? Not hardly! For, no _finite_ observer ever has seen or ever can see (or in any other way experience) either infinitely large or infinitesimally small anything—space or otherwise. When was the last time anyone _observed_ anything $10^{1,000,000}$ light years in diameter, let alone $10^{infinite}$ light years? When was the last time anyone _observed_ anything 10^{-100} cm. in diameter, let alone $10^{-infinite}$ cm.? What, then, shall we say to those who claim either that space is infinitely extended or that space and its occupants are infinitely divisible or both?

This: Though they are apparently oblivious of it, their assertions are ones they cannot possibly confirm by _observation_ and, consequently, cannot _rightly_ call _scientific_ or _empirical_ _facts_. On the face of it, their assertions of infinite this and infinite that are merely _inferences_ they _speculate_ are implied by what they experience. I say "speculate", because they can't confirm by _observation_ whether what's experienced implies the inference wholly on it own or under the influence of what they need it to imply. They, though, remain oblivious of that and, thus, can't see how manifestly wrong it is to call infinite extension and infinite divisibility observable facts rather than pure conjecture regarding what is _implied_ (and how implied) by what's observed—implied to _them_, that is.[2]

To _me_, the observable facts do not imply anything in this universe is in any way infinite. That's because, for me (as for, at least, Zeno of Elea and Leibniz), reason quickly tells me the mind is confronted with mind boggling absurdities the instant anything in this universe is said to be in any way infinite.

First, we must ask if—as most philosophers of yore concluded—some kind of infinity is found in material bodies. Or, is that impossible? Whichever way you answer that question, it's by no means _un_important. On the contrary, it's _all-important_ to our quest for truth. That's because this particular issue has almost always been the chief cause of the feuds among those who have written about nature. That's the way it _has_ been and _must_ be. For, no matter how little you stray from the

---

[1] Not everyone views space as pure nothingness. For some, "time-space" is some kind of field. For myself, it remains to be seen in what sense a "time-space field" is a _reality_.

[2] Some will object: "Many observed facts so strongly imply infinite space and infinitely divisible space, spatial objects, locomotion, and time, none can sanely question the inference." The same was wrongly said of Geocentrism by virtually every human being on Earth for thousands of years.

truth at the start, it winds up a thousand times worse down the line.
——ARISTOTLE: *On The Heavens*; 271^b, 2-9. My rather free translation - ENH. Compare mine with the translation given at the middle of the right-hand column of page 362 in volume 8 of **Great Books Of The Western World** as published by William Benton, Chicago, 1952.

How right you are, Aristotle! From the beginning, infinite divisibility has been "the chief cause of the feuds among those who have written about nature."

I repeat: Infinite divisibility is not an observable fact; it is strictly an inference *supposedly* implied by the images our senses present to us. It, therefore, needs corroboration. Furthermore, from at least as early as Zeno of Elea (c495-c430 BC), protestors have brought formidable arguments against the notion of infinite divisibility (*a/k/a:* the notion of the continuum). To be sure, all those protestors ever did was to point out the mind-boggling absurdities inherent to the concept of infinite divisibility. At no time did they put forth the slightest trace of an explanation of how what we observe could appear infinitely divisible without actually being such. With their words, they could destroy the notion of a continuum but could not even begin to build anything in its place. Oppositely, Esoptrics—for the first time in history—puts forth a highly detailed and mathematically precise explanation of how what we observe can appear infinitely divisible without being such. That explanation boils down to history's first "picture" of what's *logically* rather than *physically* extended and, thus, of how a single point is *logically* divisible into an astronomical number of points *logically* outside of one another in the order of sequence. Sadly, it may be I alone can make any sense out of *logically* versus *physically* extended.

Combine the arguments of Zeno et al with Esoptrics' explanation, and the evidence against infinite divisibility is so formidable, it becomes an inference requiring an *ocean* of corroborating evidence—an ocean so vast, it's beyond anyone's ability to amass it. As a result, the notion of infinite divisibility cannot *rightly* be regarded as anything but pure conjecture, until such time as someone can actually observe infinitesimal segments of space, spatial objects, time, and locomotion. But how is it even remotely possible for a *finite* observer to do that?

Decades ago, my opposition to the notion that "some kind of infinity is found in material bodies" (or in space, or time, or locomotion, for that matter) soon enough led me to Esoptrics' contention that no segment of either space or a material body is *smaller* than roughly 10^{-47} cm. or *larger* than roughly 10^{31} cm., which is to say 2^{-129} vs. 2^{128} respectively times the diameter of the hydrogen atom with the electron in its outermost orbit. In case you're wondering, $2^{128} = 3.4 \times 10^{38}$.

Why such limits?! Esoptrics speaks in the terms of *primary acts* (See par. 3 pg. 3) instead of *bodies*. You might say its thinking (as is St. Thomas Aquinas') is *act*- rather than *body*-oriented. That's because, for Esoptrics, no finite being can *picture* anything whatsoever about the internal characteristics of the universe's *ultimate* constituents *as bodies*. No, that doesn't make them *wholly* un<u>knowable</u>—just *wholly* un<u>imaginable</u>. Just as one born blind cannot be aware of what it's like to experience colored shapes, no one this side of infinity can be aware of what it's like to experience the internal characteristics of an ultimate constituent of the universe. That's why Esoptrics says the universe's *ultimate* bodies are bodies in a sense we the living cannot even begin to grasp. As a result, Esoptrics says the mathematical implications of their primary acts are all we the living can know about the *ultimate* "bodies" (in parentheses because they are bodies in no sense we can fathom).

The first part of those implications deals with the relative powerfulness of some *ultimate* "body's" primary act. For Esoptrics, the *most* powerful primary act in the universe is 2^{256} (*i.e.:* 1.15792 x 10^{77}) times as powerful as the *least* powerful; consequently, if the most powerful act extends its influence throughout a segment of space 1.7 x 10^{31} cm. in diameter (*i.e.:* roughly 18,000,000,000,000 light years), then the least powerful act extends its influence throughout a segment of space only 7.3 x 10^{-47} cm. in diameter.

But, the above reference to *spatial diameters* is merely a concession to the way the senses virtually *force* body-oriented minds to speak. Speaking to act-oriented minds, Esoptrics would say: The most powerful act extends its influence throughout every one of the 2^{256} levels of power, and the least powerful act extends its influence only to the lowest of those 2^{256} levels. To clarify what that implies, say to Esoptrics: "What lies beyond the outermost limits of space?" Esoptrics will reply: "An unavailable level of power higher than the highest level currently available to creation!" Likewise, say to Esoptrics: "What lies below 10^{-47} cm.?" Esoptrics will reply: "An unavailable level of power lower than the lowest level currently available to creation!"

I repeat: For Esoptrics, physically extended space and spatial objects are merely erroneous inferences mistakenly viewed as observable facts. That means this: For Esoptrics, the entire universe is *physically* a singularity: It's a single point having no *physical* length, breadth, depth, or what have you. How, then, is it possible to have different locations and parts of spatial objects *outside* of one another?

B-3. *LOGICALLY* EXTENDED SPACE & SPATIAL OBJECTS:

For Esoptrics, different locations and parts of spatial objects are *logically* outside of one another *in the order of sequence*. That "order of sequence" is one which governs both: (1) the various kinds of primary acts which the ultimate bodies can perform, and (2) the way they can change from one primary act to another. How can that be? To answer that, let's first describe the simplest way in which the various primary acts are logically outside of one another.

Hereafter, let the symbol Z designate the power of the *most* powerful primary act. That done, we can now say that the power of the *least* powerful one is 2^{-256} Z. Next, Esoptrics says there are—throughout what it calls this the universe's 9th epoch—neither more nor less than 2^{256} (*i.e.:* 1.15792 x 10^{77}) different levels of power. The least = 2^{-256} Z; the next = 2(2^{-256} Z); the next = 3(2^{-256} Z); the next = 4(2^{-256} Z); the next = 5(2^{-256} Z); the next = 6(2^{-256} Z); the next = 7(2^{-256} Z); etc.. For convenience's sake, let's refer to these levels as levels OD1, OD2, OD3, OD4, OD5, OD6, OD7, etc., respectively.

In Esoptrics, "OD" is short for "ontological distance" and "distance from zero Being". It's one of Esoptrics' most important terms and always designates which of the 2^{256} levels of power a particular ultimate is currently employing.

As a result, a single *physically* dimensionless point (*i.e.:* single singularity) can contain 2^{256} "bodies", each of which is *logically* distinct from, and *logically* outside of, all of the others by virtue of the fact that the power of its primary act is unique to itself. Since such bodies are *logically* distinct and separate, they do not need to be *physically* outside of one another.

If that's difficult for one to accept, I dare suggest it's because one is trying to think of *ultimate* "bodies" in the terms of the kinds of bodies we sense. One's mind is thus body-oriented in the most primitive sense of "body" (*viz.:* an inherently 3 dimensional lump of stuff so perfectly homogeneous, no portion of it fails to be the same, one kind of stuff no matter *how small* that portion might be). As some might say it: One's mind is still too "carnal", and carnal minds can't understand that *ultimate* "bodies" are so different from *sensible* bodies, they, unlike the latter, can be logically distinct and separate from one another as a result of nothing more than the differences in the primary acts they are currently performing.[1]

But, what does "in the order of sequence" mean? To answer, let's lay down a law which says this: As long as it does not go below the level of 2^{-256} Z or above the level of Z, each of these ultimate "bodies" can either decrease or increase its level of power to some extent; but, in doing so, it must not ***skip*** any one of the given 2^{256} levels of power. For example, suppose an *ultimate* "body" is performing OD1's act, but is accelerated to the act characteristic of OD2^{128} (*i.e.:* to the act whose level of power is $2^{128}[2^{-256}$ Z$] = 2^{-128}$ Z). Doing so, it must perform each of 2^{128} different acts in their logical sequence. For example, its act at OD1 is first in the order of sequence; next, its act at OD2 is second in that order; next, its act at OD3 is third in that order; and so forth. Each of its acts is thus *logically* outside of all of the others *in the order of sequence*. As a result, while this ultimate "body" is still at OD1, OD2 is 1 act "outside" of it; OD3 is 2 acts "outside" of it; OD4 is 3 acts "outside" of it; etc.; and yet, all are in the same, one, *physically* dimensionless point.

Here, I should perhaps clarify one issue in passing: The rate at which the ultimates go thru the logical sequence of acts varies astronomically; but, they all obey the sequence. Some ultimates may go thru a given sequence 2^{256} before others go thru it once; nevertheless, they all go thru the sequence. In a manner of speaking, all the "frogs" in the pond must land on every one of the lily pads in the line of pads; but, some of them rest on each lily pad a lot longer than others do. In Esoptrics, that's what explains how locomotion admits of various velocities.

Back on the track! In order to preserve logical separation, we must require this: (1) as an accelerating ultimate acts at OD2, the ultimate previously at OD2

---

[1] One of its most impressive achievements is the way Esoptrics explains the relationship between our sense images and the universe's *ultimate* "bodies". For Esoptrics, every *ultimate* "body" is either a "piggyback form" or a "carrying generator", and every single instance of the former is eternally concentric to some single instance of the latter. Our sense images are lines of force (*i.e.:* units of tension) generated whenever a "carrying generator" performs a particular act. For, in each of its acts, every generator acts in the presence of the drag inducing differential between its actuality and its potentiality. Our sense images are thus a kind of transitory by-product ("excretion"?) of the acts of the generators. As such, our sense images are a kind of parasitic quasi-reality and show us only the internal characteristics of a parasitic "excretion" and absolutely nothing of the *internal* characteristics of the universe's non-parasitic *ultimate* "bodies" *as actual bodies and fully real realities*.

must decelerate to OD1; (2) as an accelerating ultimate acts at OD3, the ultimate previously at OD3 must decelerate to OD2; and so forth. It's a way of saying this: No 2 ultimates can simultaneously perform the same, one act.

Hoping it will make it easier for my reader to follow what I'm saying, let me offer my reader a *figurative* explanation by means of a horizontal line like so:

0  A

Imagine the line is a meter (*i.e.:* 39.37 inches) in length and, therefore, 1,000 millimeters (*i.e.:* 1mm = .03937 inches). The line thus *figuratively* represents a universe in which there are only 1,000 different levels of power.

As we look at the black line, the left-most 1 mm. of the line suddenly shines green (or red or blue or yellow or whatever you prefer, since this is merely an allegory). It's a *figurative* way of saying a first ultimate is developing only a thousandth of the power available to it, which is to say it's actualizing only the lowest 1 of the 1,000 levels of available power. As we continue looking, the left-most 2 mm. of the line suddenly shine green. It's a *figurative* way of saying a second ultimate is developing two thousandths of the power available to it, which is to say it's actualizing only the second lowest of the 1,000 levels of available power. The first ultimate is still there developing a thousandth of the power. It is to no extent either removed or supplanted by the second ultimate, because—though the two are not outside of one another in *space* (a fact of no consequence to them)—they are outside of one another from the standpoint of the two different levels of power they are developing. In other words: They are not *physically* separate, but are *logically* so.[1]

As we continue to watch, the left-most 3 mm. of the line suddenly shine green, then the left-most 4 mm., then the left-most 5 mm., and so on, until the entire line is green. That now gives us a thousand ultimates all on the same one line; and yet, though no one of them is *physically* outside of any of the others, each is *logically* outside of every one of the others because of the fact that each is developing (*i.e.:* actualizing) a discrete percentage of the power available to them all.

If you've got that picture clear in your head, then perhaps you can notice this aspect of our *figurative* explanation: No two of our ultimates are logically *concentric* with one another. We shall soon remedy that after we first pause to consider 2 other issues.

Go back to the original situation in which only the left-most 1 mm. of the line is green. Suppose that's the only ultimate around. Can it *suddenly* cause the entire line to shine green? No, it cannot. To achieve such a goal, it must first cause the left-most 2 mm. to shine green, and then cause the left-most 3 mm. to shine green, and then cause the left-most 4 mm. to shine green, and so forth. Thus, while it's actualizing a thousandth of the power available to it, it is 999 steps (*i.e.:* locations, if we speak figuratively) from its objective. In short, 100%0A is 999 steps logically outside of it *in the order of sequence*, just as, if you must count by 1's, the number 10 is 9 steps logically outside of the number 1 in the order of sequence imposed by

---

[1] Were the second ultimate a conscious one, the first ultimate would seem to the second one to be a different ultimate half the second ultimate's own size and underlying it.

the need to count by 1's.

Go back to the situation in which a first ultimate is actualizing the left-most 1 mm. while a second is actualizing the left-most 2 mm., and let's imagine they are the only 2 ultimates around. Can the first ultimate suddenly commence to actualize the left-most 2 mm? No, it cannot do so, unless, of course, the second ultimate either accelerates and commences to actualize the left-most 3 mm. or decelerates and commences to actualize the left-most 1 mm.. It's a way of saying no two ultimates can simultaneously actualize the same power.

Now, the fun really starts. Esoptrics says there is not merely one power to actualize. After all, Esoptrics is the logic of the *mirror*; and so, the power represented by line 0A must have a mirrored (*i.e.:* anti-) power which we shall call B0. We can now give a figurative explanation of that by means of a horizontal line like so:

B _____ 0 _____ A

Here again, line 0A is 1,000 mm. in length and represents a universe in which there are only 1,000 different levels of *power*. Line B0 is also 1,000 mm. in length and represents a universe in which there are only 1,000 different levels of *anti*-power. The whole line, B0A thus represents a universe in which there are only 2,000 levels of either power or anti-power.

We can now graphically give a *figurative* explanation of two important factors: (1) how anywhere from 2 to 2^{256} different ultimates can be logically concentric with one another, and (2) how the universe's ultimates first fall into two major categories called: "piggyback forms" and "carrying generators".

Whenever a *form* actualizes a certain percentage of the *power* available to it, it simultaneously actualizes that same percentage of the *anti*-power available to it. To illustrate that, imagine this: The *left*-most 1 mm. of line 0A suddenly commences to shine green. Simultaneously, the *right*-most 1 mm. of line B0 suddenly commences to shine green. That's a *first* ultimate developing only a thousandth of both the power and the anti-power available to it.

As we continue looking, the left-most 2 mm. of line 0A suddenly shine green, and, simultaneously, the right-most 2 mm. of line B0 commence to shine green. That's a *second* ultimate developing two thousandths of both the power and the anti-power available to it. The first ultimate is still there developing a thousandth of the power/anti-power combo available to it. It is to no extent either removed or incorporated into the second ultimate, because—though the two are not outside of one another in *space* (a fact of no consequence to them)—they are outside of one another from the standpoint of the two different levels of power/anti-power they are developing. In other words: They are not *physically* separate, but are *logically* so.

As we continue to watch, the left-most 3 mm. of 0A and the right-most 3 mm. of B0 suddenly shine green, then the left- and right-most 4 mm, then the left- and right-most 5 mm., and so on, until the entire 2 lines are green. That now gives us a thousand ultimates all on the same one line; and yet, though no one of them is *physically* outside of any of the others, each is *logically* outside of every one of the others because of the fact that each is developing (*i.e.:* actualizing) a discrete percentage of the power available to them all. It also gives us a thousand ultimates which are logically *concentric*.[1]

Such is part of the way piggyback forms operate in Esoptrics' universe. I cannot here go into what happens when they accelerate or decelerate. Nor can I here take the time to explain how and why they rotate. Let's now turn to the generators.

What makes them possible is this: Whereas one class of Esoptrics' ultimates (*i.e.:* the forms) always actualize—and equally so—both the power and the anti-power available to them, another class (*i.e.:* the generators) can never simultaneously actualize both a power and its anti-power. Furthermore, whether actualizing a power or an anti-power, every generator always actualizes the same, one, highly limited percentage of the power/anti-power combo available to it. In other words: Speaking figuratively, if the power/anti-power combo be taken as a line B0A 2 meters long (*i.e.:* 2,000 mm.), then every generator always actualizes only 1 mm. per act, which is to say .0005 = .05% of the whole power/anti-power combo.

To use graphics to give a *figurative* explanation of what that means, do this: Go back to the line B0A on the prior page, and imagine this: The left-most 1 mm. of line 0A suddenly shines green; but, no part of line B0 does. That's a generator actualizing a thousandth of the *positive* power available to it. It now tries to actualize 2 thousandths of the positive power available to it, which is to say it tries to make the left-most 2 mm. of 0A shine. As it tries to do so, the line B0 simultaneously grows another 1 mm. in length in the direction of point A. It thus turns what was the left-most 1 mm. of 0A into the right-most 1 mm. of line B0. As a result, the generator still actualizes (*i.e.:* causes to shine green) the left-most 1 mm. of line 0A; but, it's now actualizing the left-most 1 mm. of a line which is only 999 mm. in length. To illustrate that, merely go back to line B0A and move 0 a single mm. to the right, and then imagine the left-most 1 mm of the shortened line 0A shining green.

Naturally, the generator may turn back to the left. As it does so, line B0 *contracts* 1 mm. to the left, and line 0A *expands* 1 mm. to the left. In its next act, the generator may continue to the left and, thereby, cause the right-most 1 mm. of line B0 to shine green. In its next act, the generator may continue its journey to the left. As it does so, line 0A *expands* 1 mm. to the left as line B0 *contracts* 1 mm. to the left. It's all a way of figuratively explaining the fact that every power and its anti-power are convertible one into the other.

But, what makes a generator a *carrying* generator and every piggyback form a *piggyback* form? Every form not an angelic one is always logically concentric with a particular generator to which it is forever joined. As a result, as the generator moves left or right along the line B0A, the center of the generator's piggyback form always is wherever the center of its carrying generator is.

How can that be? Are they not both actualizing the same one line B0A—the piggyback form in one manner, and its carrying generator in another manner? No they are not. Every form always actualizes a power/anti-power combo made *directly*

---

[1] If you truly follow what's just been said above, then I dare say you must eventually come to the realization that Esoptrics is the only cosmological theory which explains how black holes are possible, which is to say it alone explains how you can have a *point* (*i.e.:* a singularity) in the universe which is astronomically more massive than are most other points. For, it alone explains how there can be as many as 2^{256} ultimates concentric with one another at the same one logically separate *point*.

available to it by The Infinite. Oppositely, every generator actualizes a power/anti-power combo made _in_directly available to it by The Infinite by means of a form which is other than the generator's piggyback form. I'll not elaborate here.

As even carnal minds can probably see, the above presents us with a singularity (*i.e.:* a *physically* dimensionless point) which is only *one* dimensional. How does Esoptrics account for a *three* dimensional singularity? Esoptrics' answer is one very difficult either to understand or to accept. For, it says this: Because of the influence of an intrinsically triune, *infinite* "body", there are *six* kinds of power made available to every *finite* ultimate "body". More precisely, there are *three sets*, each of which includes a power and its anti-power. Each of these 3 sets (*i.e.:* "3 power/anti-power combos", if you prefer) is equally related to the other 2.

That's a logical relationship we can easily express figuratively by means of a drawing. To do that, we merely take 3 instances of the line B0A, and make each perpendicular to the other two. Naturally, they cannot all be called B0A; and so, we shall continue to call one of them "line B0A" call another "line C0Aprime" (*a/k/a:* C0A'), and call the third one "line Cprime0Bprime" (*a/k/a:* C'0B'). Do that, and we then have a cube figuratively representing the relationships between the three sets of power/anti-power. In case you missed it, B is the anti-power of the power A; C is the anti-power of the power A'; and C' is the anti-power of the power B'.

Now, the result is a *physically* dimensionless point which, nevertheless, is *logically* a three dimensional universe by virtue of the logical relationships between: (1) each power and its anti-power, and (2) each set of power/anti-power and the other 2 sets. Indeed, because each of the 3 powers and each of the 3 anti-powers admits of 2^{256} different levels, the *physically* dimensionless point is *logically* an astronomically vast 3 dimensional universe in which an astronomically huge number of cubes are interrelated to one another in an astronomical number of different patterns. How can that be?

Because I have not enough skill to draw a three-dimensional figure on a flat surface, I shall have to ask my reader to be content with the drawing of a square as given on the next page. To turn it into a cube, we must use our imaginations. Imagine, then, 2 lines each equal in length to each of the lines A0, 0B, A'0, and 0C. One of the 2 is perpendicular to the front of the page at point 0, and the other is perpendicular to the rear of the page at point 0. If the page is lying flat, the latter protrudes downward into the air behind the page and from 0 to C'; the former protrudes upward into the air in front of the page and from 0 to B'. The resulting cube shall then be a large cube composed of 512 smaller cubes, which is to say $(2 \times 4)^3$.

The pictured cube is the figurative representation of a form which is at OD4 (*i.e.:* a form at that level of power describable as $4[2^{-256}Z]$) and which is presenting the 3 sets of power/anti-power to any generator operating in reference to the said form. Be sure to realize that we are not here talking about this form's *own* carrying generator. Every generator operating in reference to the said form is, to be sure, a *carrying* generator, but carries some form other than the one represented by the drawing. Esoptrics distinguishes between the form which a generator is carrying and the other form to which the generator is "in potency", which is to say the form thru which the generator is receiving its 3 sets of power/anti-power.

Each of the *little* cubes represents what carnal minds see as a place in space,

linear locomotion is possible in a universe which is *logically* astronomical in its vastness but *physically* a dimensionless point. As for how *curvilinear* locomotion comes about, that is an issue too complex to be discussed here. Then, too, what's said here about rectilinear locomotion is more than enough to show how it's possible to have *logically* extended space and spatial objects.

C. REGARDING THE SECOND SACRED COW, CONTINUOUS TIME:

Let us turn now to the issue of continuous time. In the thinking of the vast majority of even the most educated and brilliant people, time is basically the same as space—namely: Infinitely elongated at least into the future (if not into both the future and the past), there is no limit to how small the smallest segment of it is.

Here again, of course, these assertions of theirs are—most manifestly are—ones they cannot possibly confirm by observation; and so, their assertions can't be anything but inferences; and yet, for them, they are sacrosanct, utterly unassailable, foregone conclusions of which they are absolutely certain.

Oh, and how many of these same people refuse to believe in God, unless they can observe God's existence rather than rely on what merely *implies* God's existence! Without confirmation by observation, they can be absolutely certain of *infinite matter* and *infinite nothingness*, but to no extent certain of *infinite intelligence*.

In Esoptrics, no segment of time is smaller than roughly 7.2×10^{-96} sec. by our reckoning (a unit Esoptrics calls 1 alphakronon or 1K for short), and the universe, as we know it, cannot last more than roughly 18 trillion Earth years (*i.e.:* 5.675×10^{20} seconds = $2[(2^{128})^3] \times [7.2 \times 10^{-96}]$ sec. = 2^{385}K). How can there be no segment of time smaller than that? It's because, for Esoptrics, no ultimate changes from primary act to primary act more than once per 1 alphakronon. So what?!

Esoptrics says no ultimate is involved in *time* while performing the same one primary act. Time applies to it only as it goes from one such act to another. Thus, ask for how long a *time* a given ultimate performs a given primary act. Esoptrics shall insist *time* cannot answer *absolutely*, because *time* measures <u>not</u> the *duration* of any given primary act on the part of any given ultimate; time measures only the relative frequency with which any given ultimate *changes* from one primary act to another. Time thus comes in instantaneous but intermittent spurts. Rather than being continuous, time pulses, and the rate at which it pulses for any given ultimate is relative to the rate at which that ultimate goes from primary act to another. That, of course, means time can give us a *relative* answer to the above question and say: "Ultimate so-and-so changes ten times as frequently as ultimate so-and-so; and so, in each of its primary acts, the latter stops 10 times longer than the former does in each of its acts; and so, for the former, time moves 10 times as fast as it does for the latter." Say one primary act's duration is ten times that of another, and you say nothing *absolutely* of either act's *duration*. Time simply doesn't apply.

For most who hear it, the above assertion is meaningless gibberish. That's because they do not appreciate this: As long as a given *ultimate* (such as the human soul) is performing the same, one act, it undergoes no kind of internal change whatsoever, and that leaves it unable to *experience* any kind of change whatsoever. As long as that's true, the given *ultimate* has no way to determine the duration of its act, and its act (more precisely, *each* of its acts), shall, to it, thus seem to have zero duration; and yet, from the standpoint of a second *ultimate*—which is changing from act to act at an astronomically higher frequency—each act on the part of that first *ultimate* may seem to last as much as trillions, of trillions, of trillions, of trillions, of trillions, of trillions, of trillions of millennia. Here again, because all of this is very detailed and laboriously spelled out in other works of mine, it's senseless to dwell on it here.

Still, let me make this passing remark: Esoptrics makes it quite easy to tell Zeno of Elea how Achilles overtakes the tortoise: Achilles' underlying generators change from act to act more frequently than do those of the tortoise; and so, Achilles often moves while the tortoise doesn't. It's as simple as that.

D. REGARDING THE THIRD SACRED COW, CONTINUOUS LOCOMOTION:

Finally, there is this business of continuous locomotion. For example, in the view of perhaps all but this author, Earth's motions around its axis and around the Sun actually *are* absolutely smooth; consequently, there is no limit to the smallness of the smallest segment of those two motions. In other words, such supposedly uninterrupted locomotion is, for them, infinitely divisible.

Here again, when some insist all *apparently* smooth locomotion is *actually* smooth and, therefore, infinitely divisible, their assertion is one they cannot possibly confirm by observation and, consequently, can never rightly call an empirical, scientific fact. Though they see it not, their assertion is merely an inference they *assume* is implied by what they experience, while, at the same time, they remain utterly oblivious both of what they are doing and of how manifestly wrong it is to call infinitely divisible locomotion an observable fact rather than pure conjecture regarding what is implied (and how) by what's observed.

What, then, is locomotion for Esoptrics, and how is it discontinuous despite appearing continuous? For Esoptrics, every *ultimate* "place in space" (*i.e.:* one carnal minds would describe as "roughly 10^{-47} cm. in diameter") is actually a primary act which either currently is being, or can be, performed by a generator. In other words, for Esoptrics, "ultimate place in space" and "primary act on the part of a generator" are equivalent terms, and for a generator to be in a particular "place in space" is, more exactly, to perform the primary act which constitutes that "place in space". For Esoptrics, then, all locomotion is the result of one or more carrying generators changing from one primary act to another and, thereby, creating what appears to our senses as a change from one place in space to another.

But, whenever generators do such, they linger in the course of each act. As a result, all locomotion is, for Esoptrics, *stop and go* rather than *continuous*. As with the *apparently* smooth motion of the images on a motion picture screen, all continuous locomotion is actually a composite of still frames not observed as such due to the fact that the still frames change too frequently for the human eye to detect the lack of infinitely divisible movement. All *apparently smooth* locomotion is thus no more infinitely divisible than are time, or space, or spatial objects.

Let's express that allegorically and figuratively by first going back to drawing #6 on page 22. Next, imagine a generator is occupying square 25. That's the way carnal minds would express it. Esoptrics says its performing the act characteristic of square 25. Next, imagine the generator must move to square 32. Can it just cease performing the act characteristic of square 25 and immediately start performing the act characteristic of square 32? By no means! It must follow the order of sequence. Thus, first it stops performing the act characteristic of square 25 and starts performing the act characteristic of square 26; second, it stops performing square 26's act and starts square 27's act; third, it stops performing square 27's act and starts square 28's act; fourth, it stops performing square 28's act and starts square 29's act; and so forth. Its locomotion from 25 to 32 is "continuous" only in the sense that the rate at which it changes from act to act continues the same throughout the locomotion taking it from 25 to 32. As it starts that locomotion, square 26 is logically one act "outside" of it in the order of sequence; square 27 is logically two acts "outside" of it in the order of sequence; square 28 is logically 3 acts "outside" of it in the order of sequence; and square 32 is logically 7 acts outside of it in the order of sequence.

As for the opponents to such a take on locomotion, I suspect they will fall back upon a principle dear to their hearts—one from Isaac Newton which says that every body continues in its state of rest, or of uniform motion in a straight line, unless it is forced to change its state by forces exerted upon it. Esoptrics, though, merely counters: Every *ultimate* microscopic "body" (*i.e.:* every generator) continues to change from act to act (*i.e.:* "from ultimate place in space to ultimate place in space" according to carnal minds) *at its current rate* of change and in its current logical sequence (*i.e.:* "in its current direction" according to carnal minds), unless it is forced to change its rate and/or logical sequence by forces exerted upon it.

Note that, when Esoptrics speaks of locomotion, it refers only to those *ultimate* "bodies" it calls "carrying generators" and does not refer to the ones it calls "piggyback forms". The reason for that is this: For the generators, the forms are the source of the primary acts which they can perform. As carnal minds might prefer to say it: The forms are the "space envelopes and units of pixelated[1] space in which the generators can move around from place to place". As a result, when forms change from act to act, they do not by any means "move from place to place"; instead, they either expand or contract the number of acts they make available to whatever generators are in their control. As carnal minds might prefer to say it: When forms change from act to act they either increase or decrease "the diameter of the pixelated space envelope" available to the generators in their control.[2]

---

[1] Here, "pixilated" means *composed* of pixels rather than *exposing* its pixels to the naked eye. Of course, by "pixel", I mean an ultimate unit.

[2] Changing the "diameter" of a pixilated space envelope does not change the "diameter" of the pixels composing it. Nothing ever does. It only changes the *number* of pixels composing it. I put "diameter" in quotation marks, because the pixels are *logically* cubes, and cubes have length, breadth, and depth rather than diameters. If I use "diameter", it's to save paper and trees.

For Esoptrics, then, the universe—if we may speak after the manner of carnal minds—is by no means a single, infinitely extended, infinitely divisible, universally homogeneous ball of nothingness; instead, it's a collection of an astronomically huge number of pixelated space envelopes each of which—though every pixel's size is forever the same—exhibits an astronomically huge number of different sizes. As some of today's thinkers are fond of expressing it: The _uni_verse is actually a _mul_tiverse. To say each of these space envelopes (Say "membranes", if you prefer.) is pixelated is to say each presents to one or more generators (Say "strings", if you prefer.) a definite number of what carnal minds will call "places in space" but which act-oriented minds will call "primary acts arranged in the logical sequence established by the power-anti-power combo unique to each". On this issue, it's senseless to say more than that here and now.

PART II: GOALS:

Some 52 years ago (in 1957), when the above concepts first started to form in my mind, I actually thought I would be able to _corner_ those who dared to disagree with me on the issues of infinite divisibility and physical extension. That means this: I thought I would eventually produce formulas and predictions so conclusive, none would be able to deny the truthfulness of what I eventually commenced to call "Esoptrics: The Logic Of The Mirror". Now, at 73 years of age (as of April 13, 2009), I realize it's not likely I'll achieve that goal. Why?!

There are three factors which compel me to admit failure: (1) It's all too obvious that 52 years of effort have put me nowhere near the formulas and predictions I once hoped and still need to produce to prove my theory and its cosmological model; (2) at 73, I don't have much time left to go as far in that direction as I once hoped and still need to go;[1] and (3) my intellectual ability and knowledge of physics and math are too paltry to allow me to go much further in that direction in what little time I have left. Indeed, as paltry as that ability and knowledge are, it remains to be seen if all the time in the world would be enough to go as far as I once hoped and still need to go. So paltry are they, I strongly suspect it has been only by the grace of God that I have made even what little progress I have made.

These days, then, I have a greatly reduced goal. How shall I describe it? Let's try the following.

Notwithstanding Galileo's claims, the Catholic cleric Nicolaus Copernicus did not by any means _prove_ the Geocentric Theory wrong or the Heliocentric Theory right. With its circular rather than elliptical orbits, his explanation of how Earth and the other planets orbit the Sun was admittedly wrong. For all of that, despite the paucity and erroneousness of what he said, it was enough to make it sufficiently clear to Kepler and others that Geocentrism was by no means the only possible way to explain our sense images' view of the heavens. In doing that, he put

---

[1] From roughly 4PM on Friday, September 18, 2009, to roughly 5:45 PM on Monday, September 21, I was in the hospital with multiple blood clots in my lungs sprayed there by my left leg. A nurse practitioner said to me: "You broke all the rules. We see people die with fewer blood clots in their lungs than you have; and yet, you don't even have shortness of breath."

27

an end to Geocentrism's ability to stand as a sacrosanct, utterly unassailable, foregone conclusion and, thereby, initiated that sacred cow's slow but sure demise.

My goal now is to do the same with regard to continuous time, continuous locomotion, and physically extended space and spatial objects. These days I dare to hope that—as skimpy and riddled with errors as my explanation of Esoptrics might be—it shall still match modern Science's findings on enough points to make it sufficiently clear that physical extension and infinite divisibility are not the only possible way to explain our sense images' view of time, locomotion, space, and spatial objects. My hope is that, in doing that, I, like Copernicus, shall—*even without proving my system correct!*—put an end to the ability of continuous time, continuous motion, and physically extended space and spatial objects to stand as sacrosanct, utterly unassailable, and foregone conclusions, and that, I dare predict, shall initiate the slow but sure demise of those 3 sacred cows.

That alone would be no small accomplishment. For, since they obviate the need for an extra-cosmic support (whereas logical extension and finite divisibility do require such), physical extension and infinite divisibility are the cornerstones of atheistic and agnostic Science; and so, once physical extension and infinite divisibility crumble, so must *godless* Science. Then shall be born a kind of Science which can make no final sense of nature, save thru a system of thought in which the existence of an extra-cosmic, triune, Infinite One is a mathematical necessity. These days, then, my only goal is merely to create sufficient doubt as to the sacredness of those 3 sacred cows. In short, Esoptrics, I hope, sounds their death knell and commences to reclaim Science for God.

Yes, it is nowhere near to what I once hoped to achieve; but, I must now admit it's all I can *realistically* hope to achieve, unless God deigns to intervene in a most miraculous manner. Whatever God wills! For, I am convinced that, knew I as much about my self as God does, I would observe for a fact that what God wills for me is actually what I myself most truly, *truly*, _truly_, **_truly_** choose for myself if only at a level so instinctive, I have not the slightest awareness of it. Blessed be God Who watches out for me more perfectly than I can even begin to imagine.

Fundamental ideas play the most essential role in forming a physical theory. Books on physics are full of complicated mathematical formulae. But thought and ideas, not formulae, are the beginning of every physical theory.
——**ALBERT EINSTEIN & LEOPOLD INFELD:** *The Evolution Of Physics*, as found on page 277 of the sixth paperback printing as published by Simon and Schuster, New York, 1967. An Essandess Paperback. ©1938 by Albert Einstein and Leopold Infeld.

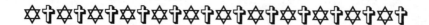

Document #2:

ESOPTRICS CORRELATES ITS ULTIMATE UNITS OF TIME & SPACE WITH LIGHT SPEED & THE QUANTITY OF LIGHT PER ACT OF CONSCIOUSNESS:

A. SOME OF ESOSPTRICS' BASIC CONCEPTS:

A-1: INTRODUCTORY:

"Esoptrics" is the name I give to the cosmological system I have been thinking about for 52 years and writing about for 47 years, as of 2009. Writing about it means I have deluged the academic, printing, and media communities with thousands of copies of letters, e-mails, essays, and books over the course of the last 47 years. Regarding books, I have, besides this one, 17 other self-published ones in print—4 of them on Esoptrics or related cosmological issues. For all of that effort on my part, there has been virtually no response whatsoever from any quarter whatsoever. I can count on the fingers of one hand (Well, maybe 2!) the number of individuals who showed any interest in my system, and their interest was extremely fleeting. To date, no one of my 4 cosmological books has sold more than 10 copies, and the 4 together have not sold 20.

The inevitable result is that not more than a handful of people have ever *heard* of Esoptrics, let alone know something about it. That being so, it would be futile of me to launch into what Esoptrics says about consciousness before I have first familiarized my reader with some of Esoptrics' basic concepts. That's what I'll now do for 27 pages and, only in the last 5, address the issue of consciousness.

"Esoptrics" comes from the Greek word for mirror. My choice to use that name stems from the fact that I look upon Esoptrics as the algebraic logic of a *mirror*. I deem it such, because it is mainly based on multiples and powers of 2, and because of its frequent recourse to reverse images. I also refer to Esoptrics as the divine algebraic logic used by God to create and maintain the universe. For, if correct, Esoptrics leaves no doubt that God exists and that the universe is a masterpiece of intelligent design.

Because Esoptrics is an *algebraic* logic, none of its principles can *really* be pictured. Esoptrics' principles are too abstract to have the picturableness of Geometry's principles. Unfortunately, human nature being what it is, it is virtually impossible for us to convey abstract principles to one another without recourse to pictures. That, then, is what we shall have to do. As we do so, just be sure to bear in mind that we are speaking *figuratively*, which is to say we—like an artist using visual shapes to convey something not picturable—are using *Geometry's* three-dimensional pictures to convey the purely abstract principles of *Algebra*.

A-2: LOGICAL VS. PHYSICAL EXTENSION & FORMS VS. GENERATORS:

Just how abstract are they? For Esoptrics, time, locomotion, consciousness, and 3 dimensional space and objects—when viewed as *continuous* and *infinitely* divisible—are—like the notion Earth is circled by the Sun and the planets—only erroneous interpretations of *ephemeral effects* which can be *sensed*. They are all phenomenal rather than noumenal, since the *enduring realities* which nothing can *sense* are all *physically* (though not *logically*) indivisible quanta. How so?!

First of all, for Esoptrics, neither space nor extension in space are *physical*; they are *logical*. Insist on speaking of the *physical*, and the entire universe is a singularity—a single dimensionless point with no trace of *physical* length, breadth, or depth. No 2 of its *ultimate* locations are *physically* outside of one another in *space*; they are only *logically* outside of one another *in the order of sequence* established and enforced by God's divine algebraic logic of the mirror. Likewise, the *ultimate* parts of a so-called "3 dimensional physical object" are not *physically* outside of one another in *space*; they are only *logically* outside of one another *in the order of sequence* established and enforced by God's divine algebraic logic of the mirror.

All of that is possible because, in reality, every so-called "place in space"—if it's an ultimate place according to Esoptrics—is, instead, a particular kind of primary act (See par. 3 on pg. 3.) which any given "generator" either currently is performing or can perform. There is no such thing as an empty place in space, save where there is a kind of primary act no "generator" is currently performing. Among these acts, there are 6 interrelated logical sequences dictating which 6 acts *must normally* follow which act, and that's what produces the purely *sensible effect* of 3 dimensional space and spatial objects. What, though, is a generator?

In Esoptrics, each of the cosmos' "ultimate building bricks"—as some relish calling them—has two halves forever logically concentric from the standpoint of each one's *actuality*. "Carrying generator" ("generator" for short) is Esoptrics' name for one of those two halves and "piggyback form" ("form" for short) is its *chief* name for the other. I emphasize *"chief"*, since I also call forms "matrices" or "space envelopes" or "secondary cubes". As with geometry's so-called "empty" space, forms are utterly permeable to one another and to generators; but, no generator is to any extent penetrable by another. Generators are adamantine; forms—like invisible mantles each surrounding a generator—are the opposite.

The name "generator" bespeaks the fact that, whenever any given one performs an act, there's a drag-inducing differential between its actuality and its potentiality, and that generates a unit of force in the sense of tension and vibration. Though not *physically* extended in *space*, every unit of force is *logically* extended *between the generator's actuality and its potentiality*. Whenever we see a visual image, hear an auditory image, etc., we're feeling what those generated units of tension are internally. Thus, our sense images do replicate a kind of extra-mental reality, but an ephemeral one parasitic unto the enduring, non-parasitic generators. As such, sense images are not so much a kind of *reality* as a kind of *quasi*-reality which must be generated over and over again one time per act of its generator.

Generators are the universe's <u>micro</u>scopic and forms its <u>macro</u>scopic ultimate constituents. That's because—if we think *figuratively*—every form can be depicted as a *permeable* (ghostly? immaterial? impalpable? ethereal?) cube whose length, breadth, and depth (LB&D for short) may vary astronomically as it expands and contracts rhythmically; whereas, every generator can be depicted as an *im*permeable (solid?) cube whose LB&D are each forever roughly $7.34683969 \times 10^{-47}$ centimeters as it "leapfrogs" its way thru the universe at one or another of an astronomical number of different frequencies.

A-3: ALPHATOPONS VS. ALPHAKRONONS & THE SPEED OF LIGHT:

Whence comes $7.34683969 \times 10^{-47}$ centimeters? One starts with the measurement of the hydrogen atom with its electron in its outermost orbit (Why shall be clear later.). That measurement is usually given as 5×10^{-8} centimeters, which is to say .00 000 005 cm., or you can write it as 1/20,000,000 of a centimeter. Next, one turns to 2^{129}. In everyone's book, that's 680,564,733,841,876,926,926,749,214,863, 536,422,912. Shorten that to: $6.805647338 \times 10^{38}$. What now is the result if we divide 5×10^{-8} cm. by 2^{129}? The result is the above $7.34683969 \times 10^{-47}$ cm.. Some, I suspect, will object to my carrying the conclusion out to 8 decimal places, saying I have no grounds for doing so. Despite that, I shall continue to do so for a reason which shall become clearer shortly. For now, let's concentrate on the fact that, for Esoptrics, $7.34683969 \times 10^{-47}$ cm. $= 2^{-129}$ times the hydrogen atom's diameter. Be sure to notice the minus sign in the exponent.

The measurement of $7.34683969 \times 10^{-47}$ cm., I call: "1 alphatopon" or "1Φ" for short. One cm. $= 1.361129468 \times 10^{46}\Phi$. For Esoptrics, there is no unit of space or occupant of space smaller than 1Φ, because no generator ever performs a basic act whose LB&D are (speaking *figuratively*) less than 1Φ.

In Esoptrics, a 2nd crucial measurement is one I call the "alphakronon" or "K" for short. $1K = 7.201789375 \times 10^{-96}$ seconds, and 1 sec. $= \underline{1.388543802 \times 10^{95}K}$. For all, $2^{128} = 3.402823669 \times 10^{38}$; and so, 1Φ per $2^{128}K = 4.080563489 \times 10^{56}\Phi$ per sec. (***i.e.:*** per $\underline{1.388543802 \times 10^{95}K}$) $= 29,979,245,802$ cm./sec.. That's about as close to light speed as Science gets and is why I carry my decimals out to 8 or 9 places.

In Esoptrics, the alphakronon, K, is crucial because: (1) It's Esoptrics' basic unit of time, and (2) Esoptrics says there's never anywhere in the cosmos any kind of change below 1K. Below 1K, everything everywhere in the universe is absolutely frozen; and so, throughout the universe, *continuous* time, space, locomotion, and consciousness are no different from what happens when, at a movie theatre, we view a motion picture's 24 frozen frames per second—namely: The continuousness of it all is a mistaken projection onto the *extra*-mental reality of a purely *intra*-mental effect. The same trick spawned the Geocentric Theory.

If you ask how long that absolutely frozen state lasts, Esoptrics does <u>not</u> answer 1K. That's because, for Esoptrics, the only kind of time we know is the kind which measures *change*; and so, 1K cannot measure *duration* in a *changeless* state.

Because one change per 1K is always and everywhere the fastest rate of change, all other rates are expressed in the terms of K, and time is relative to that rate. To be more specific: In Esoptrics, time—for any given generator or form—is relative to the rate at which that generator or form changes from act to act. More on that later! For now, hammer this home in your head: Whenever Esoptrics seems to be telling you the *duration* of some ultimate's act, it's merely telling you what that act's rate of *change* is *relative* to the universally fastest rate of change. For Esoptrics, then, there is no such thing as *continuous* or *infinitely* divisible time, just as there is no such thing as *infinitely divisible* or *infinitely extended* space.

A-4: PLACES IN SPACE VS. ACTS IN LOGICAL SEQUENCE:

I repeat: In Esoptrics, the universe is not a collection of places in space which one or more kinds of less passive realities can either occupy or traverse. Instead, it's a collection of an astronomical number of different kinds of acts which any generator can perform as—surrounded by its piggyback form—it carries that form around within the confines of a second form acting as the generator's space envelope (*i.e.:* acting as the form to which the generator is in potency).

If you want a *figurative* explanation of that, let's examine the drawing below, and imagine the squares are cubes. None are drawn to scale.

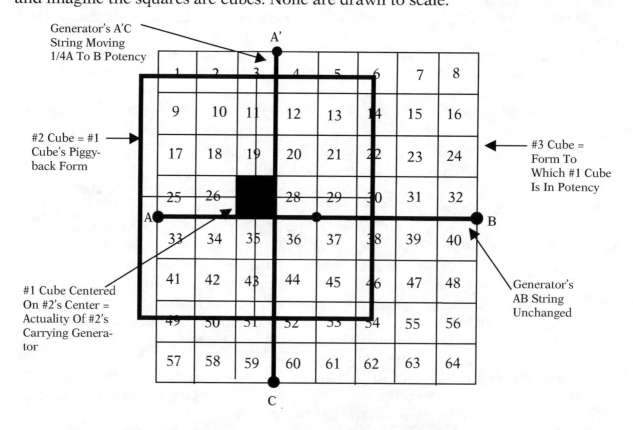

**DRAWING #7: A #2 FORM MOVING THRU A #4 FORM
BY MEANS OF THE #2 FORM'S CARRYING GENERATOR.**

32

Cube #1, a carrying generator, is an *im*permeable cube whose center is one with the center of the larger *permeable* cube #2—#1's piggyback form. With the larger permeable cube surrounding the smaller impermeable cube at its center, this combo moves around within the confines of another *permeable* cube, #3, even larger than cube #2. This larger cube, #3, is itself entirely composed (*figuratively* speaking) of 512 (*i.e.:* $[2x4]^3$) tiny permeable cubes (I can draw only 64 and only as squares.), and each of its tiny component cubes is equal to the *first* cube in LB&D. Each of cube #3's tiny component cubes is both an empty place in space (*figuratively* speaking) and a particular kind of act (Recall par. 3 pg. 3: "Act" = "Primary Act".) the *first* cube can perform. To carry its piggyback form around within cube #3's limits, cube #1 becomes one with cube #3's component cubes one per act. First, cube #1 is co-terminus with one of cube #3's tiny component cubes (#27 in the drawing) for a definite number of K (alphakronons); then it is suddenly co-terminus—for a definite number of K—with an adjacent one of cube #3's tiny cubes (in the drawing, either 19, 26, 28, or 35); then it is suddenly co-terminus—for a definite number of K—with an adjacent one; etc.. One per act, each of cube #3's tiny cubes **is** the adamantine cube #1 for some K, and *only* they are. For, as vacuous as cube #3 is, it confines cube #1 within its boundaries for as long as the latter's piggyback form (*i.e.:* cube #2) fails to undergo the required change. Be patient!

That, though, is not a case of cube #1 skipping what some would call "the infinite number of fractional intermediate places in space" as it "leaps instantaneously from one place in space to a *physically* contiguous one equal to the prior one in size"; rather, it's a case of cube #1 ceasing to perform one kind of act and commencing to perform one of the six (**viz.:** left, right, up, down, front, rear, figuratively speaking) which—as +1 & -1 are to 0—are *logically* next in the 6 fold order of sequence established by the logic of the mirror.

Newton said: "Every body continues in its state of rest, or of uniform motion in a right line, unless it is compelled to change that state by forces impressed upon it." Esoptrics says: "Every generator continues in its current one of the six orders of sequence and at its current rate of change from act to act, unless forced to change that current rate and/or current order of sequence."

Since it's very likely you didn't get a sufficient hold on all of that, let's go thru it again in more detail. As I said before, though, just be sure to bear in mind that we are speaking *figuratively*, which is to say we're using *Geometry's* 3 dimensional pictures to convey the purely abstract principles of *Algebra*. Remember too that LB&D = length, breadth, and depth. With that in mind, we recast the picture so:

First, imagine a series of 3 dimensional matrices—each a permeable cube composed entirely of smaller ones (not to be confused with the generators) each 1Φ (*i.e.:* 1 alphatopon = 7.34683969 x 10^{-47} cm.) LB&D. Each of these same-sized, tiny cubes can be called a "primary cube". As for the matrices, we can call them "secondary cubes". Of them, there are 2^{256} (*i.e.:* 1.1579 x 10^{77}) different kinds which, for now, we can name: #1 thru #2^{256}. Their numbers indicate how far they were from zero actuality at creation, and this I call their *native* state. Thus: At creation, #1 was only 1 step away from zero actuality, and that is forever its native state; #2 was only 2 steps away, and that is forever its native state; #3 was only three steps away, and that is forever its native state; #4 was only 4 steps away; etc.; until, #2^{256} is 2^{256} steps away, and that is forever its native state. For now, 2^{256} is the maximum distance from zero actuality a form (*a/k/a:* secondary cube) can be. Every form is called a matrix and space envelope because—*for one another, generators, and Geometry!*—each is a unique kind of utterly permeable "empty" space. For God and all other inhabitants of infinity, every form is very much other than any kind of void or space.

A-5: CATEGORICAL, GENERIC, & SPECIFIC FORMS:

Forms #1, #2^1, #2^2, #2^4, #2^8, #2^{16}, #2^{32}, #2^{64}, #2^{128} (*i.e.:* the complete atom's form), and #2^{256} (*i.e.:* the form of the universe) are each a *categorical* form. Surely, you notice the progression involves successive *squares* of the powers of 2. Every categorical form automatically causes its carrying generator to be in potency to the next highest categorical form. At least, that's the rule until "The Great Acceleration" produces the 8 reverse categories and, thereby, changes the rule for forms at 2^{128}. More on that later!

Every form whose number is a power of 2 is a *generic* form, unless, of course, it's one of the 10 just listed. Every generic form automatically causes its carrying generator to be in potency to the next highest generic form.

Every form whose number is neither 1 nor a power of 2 is a *specific* form. Every specific form automatically causes its carrying generator to be in potency to the next highest specific form.

Every form's native state thus determines whether it is forever a particular species or genus or category. In Esoptrics, this distinction between species, genera, and categories is super-critical.

A-6: THE ACTUALITY VS. THE POTENTIALITY OF FORMS:

One of the main differences between the various matrices is the number of *active* primary cubes per matrix. In each matrix's *native* state, that number = $(2N)^3$, where N = the matrix's number. For example, in matrix #1, the number of *active* primary cubes = $(2x1)^3$ = 8, and in matrix #2^{256}, the number of active primary cubes = $(2x2^{256})^3 = (2^{257})^3 = 2^{771} = 1.242 \times 10^{232}$. Every matrix's every *active* primary cube is one which a generator can convert into one of its own acts and, thus, is a part of some generator's *potentiality*. For those who prefer to think figuratively, every matrix's every primary cube is a place in space which a generator can occupy.

Why do I emphasize "*active*"? *Figuratively* speaking, every matrix always and everywhere is exactly $2(2^{256})\Phi$ LB&D (*i.e.:* roughly 18,000,000,000,000 light years = 1.7×10^{31} cm.).[1] That is 2^{128} x the hydrogen atom's diameter which, in turn, is 2^{129} x each generator's LB&D and 2^{128} x the LB&D of the smallest form. As calculated above, every matrix is thus composed of 2^{771} (*i.e.:* 1.242×10^{232}) primary cubes, and that allows us to say the *potentiality* of every form is forever 2^{771} primary cubes. The

[1] If memory serves me, in 1962, I was at the main library in San Antonio, TX, reading an article in the **Encyclopedia Britannica** which said that, according to Einstein, the diameter of the universe was 10^{30} cm.. I've never been able to verify that. How I snarl at myself for not making a permanent record of the edition, volume, page, etc.! Someday, maybe someone else shall find that data for me.

actuality of every form is a different story which *at first* goes like this: As long as any given matrix (*i.e.:* form) is only 1 step away from zero actuality, it actualizes only 8 (*i.e.:* the cube of 2x1) of its 2^{771} primary cubes; as long as it's only 2 steps away, it actualizes only 64 (*i.e.:* the cube of 2x2) of its 2^{771} primary cubes; as long as it's only 3 steps away, it actualizes only 216 (*i.e.* the cube of 2x3); etc.; until, at 2^{256} steps away from zero actuality, the matrix actualizes all 2^{771} of its primary cubes.

A-7: THE PULSING OF FORMS & THE NUMBER PER NATIVE OD:

Why—in the prior paragraph—did I emphasize the phrase *"at first"*? That's because every form expands and contracts at regular intervals. Each has a pulse rate just as the human heart does. As with the human heart, the result is a two-phased cycle composed of a state of contracted actuality followed by a state of expanded actuality followed by a state of contracted actuality, etc.. If a form is still at the distance from zero actuality at which it was created (*i.e.:* its native state), then, in the second half of its two phase cycle, it doubles its "size" and multiplies by 8, the number of primary cubes it's actualizing. By way of example, if a form was created at, and still at, 1 step away from zero actuality, then for 2K (*i.e.:* 2 alphak-ronons), its actuality is 2Φ (*i.e.:* 2 alphatopons) LB&D, and it actualizes 8 of its 2^{771} primary cubes; for the next 2K, its actuality is 4Φ LB&D, and it actualizes 64 of its 2^{771} primary cubes; and, for the next 2K, it's back to the starting condition. Again, if a form was created at, and still at, 2 steps away from zero actuality, then, for 4K (*i.e.:* 2^2K), its actuality is 4Φ LB&D, and it actualizes 64 of its 2^{771} primary cubes; for the next 4K, its actuality is 8Φ LB&D, and it actualizes 512 of its 2^{771} primary cubes; and, for the next 4K, it's back to the starting condition. Again, if a form was created at, and still at, 3 steps away from zero actuality, then, for 9K (*i.e.:* 3^2K) its actuality is 6Φ LB&D, and it actualizes 216 of its 2^{771} primary cubes; for the next 9K, its actuality is 12Φ LB&D, and it actualizes 1,728 of its 2^{771} primary cubes; and, for the next 9K, it's back to the starting state. It's what causes some generators to move in waves as they try to stay confined within the vacillating outer limits of the actuality of the form to which they are in potency.

The calculations become a bit more complicated when a form moves away from its native state's distance. For, in that case, in the *expanded* actuality phase, the LB&D of the form's actuality increase by only $2N_n$, where "N_n" equals the number of the distance from zero actuality at which it was created. For example, imagine the form created 1 step from zero accelerates to 2^{128} steps from zero—its acceleration limit. In that case: Throughout the *first* half of its 2 phase cycle, its actuality is $2(2^{128}) = 2^{129}\Phi$ LB&D for 2^{128}K, and it actualizes 2^{387} (*i.e.:* $[2x2^{128}]^3$) of its 2^{771} primary cubes. Throughout the *second* half of its 2 phase cycle, its actuality is $2(2^{128}+1) = 2^{129}+2\Phi$ LB&D for 2^{128}K, and it actualizes $(2^{129}+2)^3$ of its 2^{771} primary cubes. For every form at all times everywhere, its actuality's LB&D in Φ equals—throughout the 1st half of its 2 phase cycle—2 x its *current* distance from zero actuality and—throughout the 2nd half of its 2 phase cycle—equals 2 x the sum of its *current* + its *original* distance from zero actuality, even where both distances are the same.

The contents of the last 4 paragraphs can be expressed in the following six

ESOPTRICS: LOGIC OF THE MIRROR

formulas:

$$\textbf{FORMULA \#1: } \Phi_1 = 2(D_c)$$
$$\textbf{FORMULA \#2: } \Phi_2 = 2(D_c + D_n)$$
$$\textbf{FORMULA \#3: } A_1 = [2(D_c)]^3$$
$$\textbf{FORMULA \#4: } A_2 = [2(D_c + D_n)]^3$$
$$\textbf{FORMULA \#5: } K_{nc} = 2[(D_n)^2]$$
$$\textbf{FORMULA \#6: } K_{ac} = 2(D_c \times D_n)$$

In each formula, D_c = current and D_n = native distance from zero actuality.

In formula #1, the symbol Φ_1 tells us we're calculating the LB&D of any given matrix's actuality in the course of the _first_ of the 2 acts in each of its cycles. That "size", it says, equals, in alphatopons, 2 times that matrix's current distance from zero actuality.

In formula #2, the symbol Φ_2 tells us we're calculating the LB&D of any given matrix's actuality in the course of the _second_ of the 2 acts in each of its cycles. That "size", it says, equals, in alphatopons, 2 times the sum of that matrix's current and native distances from zero actuality. For example, suppose we're dealing with a matrix whose _current_ and _native_ distance from zero actuality is 2^{128}. In that case, in its first act, its actuality's LB&D are each $2(2^{128}) = 2^{129}\Phi$, and, in its second act, that changes to $2(2^{128} + 2^{128}) = 2(2^{129}) = 2^{130}\Phi$. Next, suppose the same matrix accelerates to 2^{255}. In that case, in its first act, its actuality's LB&D are each $2(2^{255}) = 2^{256}\Phi$. In its second act, that changes to $2(2^{255} + 2^{128})\Phi$. As you yourself can perhaps readily calculate, the second act involves an astronomically insignificant change.

Apply formula #1 correctly, and you find this: In the first act of a form native to, and currently at, 1 step from zero actuality, its LB&D are each 2Φ, which is to say 2^{-128} times the hydrogen diameter. Turn to the opposite end of the scale. In the first act of a form native to, and currently at, 2^{256} steps from zero actuality, its LB&D are each $2^{257}\Phi$, which is to say 2^{128} times the hydrogen diameter. That's 2^{128} versus 2^{-128} times the hydrogen diameter—the latter being $2(2^{128}) = 2^{129}\Phi$.

That's why, for Esoptrics, the hydrogen atom is what Esoptrics calls the "mirror threshold". What's below that threshold in distance from zero actuality is _inside_ the mirror, and what is above that threshold in distance from zero actuality is _outside_ the mirror, and one is the mirrored image of the other just as $\frac{1}{2}^{128}$ and $2^{128}/1$ are mirrored images of one another. I strongly suspect that the existence of the mirror threshold is what accounts for the clash between Relativity and Quantum Mechanics.

In formula #3, the symbol A_1 refers us to the actuality of a form (_i.e.:_ secondary cube/matrix) in the course of the _first_ of the 2 acts in each of its cycles. It tells us that the number of primary cubes (_i.e.:_ smaller cubes comprising the larger, secondary cube) involved in that actuality equals the cube of two times the form's current distance from zero actuality.

In formula #4, the symbol A_2 refers us to the actuality of a form in the course of the _second_ of the 2 acts in each of its cycles, and it tells us that the number of primary cubes involved in that actuality equals the cube of 2 times the sum of the form's current distance from zero actuality plus its native distance. For example, suppose we're dealing with a matrix whose *current* and *native* distance from zero actuality is 2^{128}. In that case, in its first act, the number of primary cubes comprising its actuality $= [2(2^{128})]^3 = (2^{129})^3 = 2^{387}$, and, in its second act, that changes to $[2(2^{128} + 2^{128})]^3 = [2(2^{129})]^3 = (2^{130})^3 = 2^{390}$. Next, suppose the same matrix accelerates to 2^{255}. In that case, in its first act, the number of primary cubes comprising its actuality $= [2(2^{255})]^3 = 2^{768}$. In its second act, that changes to $[2(2^{255} + 2^{128})]^3 = (2^{256} + 2^{129})^3$.

In formula #5, the symbol K_{nc} refers us to the duration in alphakronons of a complete cycle on the part of a form currently at its native state. For example, for a form currently at its native state of 2^{128}, each one of its cycles endures for $2[(2^{128})^2]K = 2(2^{256})K = 2^{257}K$. Since each cycle involves 2 acts of equal duration, the duration of each act in the cycle would be $2^{256}K$.

In formula #6, the symbol K_{ac} refers us to the duration in alphakronons of a complete cycle on the part of a form currently at a distance from zero other than that of its native state. For example, suppose we're dealing with a matrix whose current and native distance from zero actuality is 2^{128}. In that case, the duration of each of its cycles $= 2[(2^{128})^2] = 2(2^{256}) = 2^{257}K$. Suppose that same form then accelerates to 2^{256}.[1] In that case, the duration of each of its cycles $= 2(2^{256} \times 2^{128}) = 2(2^{384}) = 2^{385}K = 7.88 \times 10^{115}K$. For Esoptrics, that latter figure is the duration of our universe before another big bang occurs to produce a universe composed of $(2^{256})^2 = 2^{512}$ kinds of piggyback forms ranging from roughly 10^{-94} to 10^{62} cm. LB&D.

The larger is the number of the matrix, the fewer of that kind there are in the universe and vice versa. For example, there is only 1 instance of matrix #2^{256}, but 6^{256} (*i.e.:* 1.6096×10^{199}) instances of matrix #1. There are 6^{255} instances each of matrices #2 and #3. There are 6^{254} instances each of matrices #4, #5, #6, and #7. There are 6^{253} instances each of matrices #8, #9, #10, #11, #12, #13, #14, and #15. There are 6^{252} instances of each of the matrices from #16 thru and including #31, etc.. I assume you catch the nature of the progression and the fact it's dealing with decreasing powers of 6 and increasing powers of 2.

A-8: ACCCELERATION ON THE PART OF A FORM:

What is meant by acceleration? If two or more generators operating in potency to the same form try to perform the same, one act simultaneously, they collide *logically* and greatly affect one another's rate of change from act to act. In turn, that shall affect each of the piggyback forms riding on the colliding generators. For example: If one of the generators <u>in</u>creases its *rate of change*, its piggyback form <u>in</u>creases its *current distance from zero actuality*, and that <u>de</u>creases the form's *rate of change*. Thus, as time and velocity speed up for any generator, time slows down for its piggyback form and vice versa. Sound familiar, Dr. Einstein?

As the latter increases its current distance from zero actuality, it "swallows" one form for each of the distances from zero thru which it passes. For example, if a piggyback form native to 2 steps from zero accelerates to 2^{128} steps from zero, it

---

[1] Actually, it can't accelerate beyond $2^{256} -1$. But, to keep the example simple, let's assume it can.

will draw in and make concentric to itself, 2^{128}–2 forms together with their carrying generators (omitting the #1 matrix and its carrier since all #1's fuse only with their own kind and, doing so, produce electrons.), and each of those ingested forms shall be at a discrete distance from zero actuality. The result shall be a *super* matrix we can *figuratively* depict as a nest of boxes. The largest box contains the 2nd largest box which, in turn, contains the 3rd largest which, in turn, contains the 4th largest which, in turn, contains the 5th, and so on. Also try to imagine every box's center is the same as that of all the others. To do that, you must imagine each box fixedly floating inside the next larger one. See drawing #5 on page 13 or #5b on page 49.

The result of such a fusion is a *super* matrix. Super matrices are composed of up to 2^{256}–2 concentric matrices. There can be 6 such with anywhere from 2^{255} to 2^{256}–2 matrices and they anywhere in the universe, which is to say their center can be anywhere within the confines of the one form native to OD2^{256}. There's an astronomical number of super matrices containing matrices #2 thru #2^{128}–2 circled by a super matrix composed of up to 2^{128}–1 #1 matrices. Each of these is a hydrogen atom. Remember that 2^{128} is the square root of 2^{256}.

If a super matrix is composed of considerably more than 2^{128} matrices, it serves as the center of either a heavenly body or a galaxy or a group of galaxies or a cluster of galaxy groups, etc.. Every such super matrix can be called a *cosmic* super matrix (black hole?), and each exerts a gravitational pull proportional to the number of matrices fused together at its center. Every such *cosmic* super matrix thus causes other matrices and super matrices either to cluster *around* it or *orbit* it as opposed to becoming *concentric* with it.

A-9: THE EIGHT REVERSE CATEGORIES PRODUCED BY THE GREAT EIGHT-FOLD ACCCELERATION:

The most important acceleration is what Esoptrics calls: "The Great Eight-Fold Acceleration And The Formation Of The Eight Reverse Categories". In the following explanation, "RC" shall be short for "reverse category", "RC matrix" short for "reverse category cosmic super matrix", and "OD" short for "ontological distance" and "distance from zero actuality".

RC #1: A maximum of 6 forms native to 2^{128} steps from zero actuality accelerate to somewhere between OD2^{255} and OD2^{256}–1 thus producing 6 instances of the first reverse category. Each is a cosmic super matrix in which anywhere from 2^{255}-1 to 2^{256}-2 matrices are logically concentric. As a category, each encompasses the ontological distances of 2^{255} thru 2^{256}–1. As a result, each can be *figuratively* described as a nest of roughly 10^{77} concentric boxes/cubes in which the smallest is roughly 10^{-47} cm. LB&D and the largest anywhere from roughly 9 (at OD2^{255}+1) to 18 (at OD2^{256}-1) trillion light years LB&D.

The carrying generator of each of these six is in potency to the matrix at OD2^{256} (*i.e.:* the form of the universe) and, so, for each of the six, locomotion means using that form's 2^{771} primary cubes 1 per act. As they do so, each of the 6 RC#1 matrices moves toward a different side of the form of the universe. They do so because super matrices maximally repel one another, if the leading form of each is native to the same categorical OD and has been so accelerated as to move within the confines of the same category—in this case, the reverse category running from $OD2^{255}$ to $OD2^{256}$-1. As they move away from the center of the form of the universe, each of the 6 can travel at least 9 trillion light years before they can go no further from the center.[1]

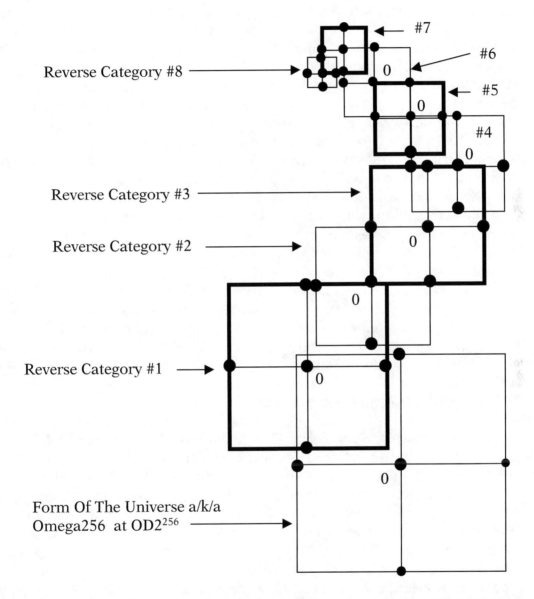

Reverse Category #8

Reverse Category #3

Reverse Category #2

Reverse Category #1

Form Of The Universe a/k/a
Omega256 at $OD2^{256}$

DRAWING #8: ONE CHAIN POSSIBLE TO THE REVERSE CATEGORIES.

In the above drawing, one of the 6 instances of RC#1 (I most certainly can't draw all 6.) is shown as having gone most of the distance from the center of the

---

[1] Why do I say "at least"? Remember: Start at the center of the cube and draw a radius *perpendicular* to one of its six sides, and, in this case, the radius shall be 9 trillion light years in length. But, start at the center and draw a radius *at an angle* to any side or corner, and the radius shall be more than 9 trillion light years in length. In every sphere, all the radii are equal in length. Not so in cubes!

form of the universe to the corner formed by the junction of the A and A' sides of the form of the universe.

I should not have to point out that the size of each square in the drawing is irrelevant. That's because there is no way I can maintain the required scale on paper this size. That should be obvious, if you bear in mind that, to maintain that scale, the LB&D of the square representing the form of the universe would have to be at least 16 times the LB&D of the square representing RC#4, at least 256 times the LB&D of the square representing RC#5, at least 65,536 times the LB&D of the square representing RC#6, and so forth.

RC#2: In RC#1, we have *one* RC#1 matrix moving toward each of the 6 sides of the next highest categorical form—the latter being the one form native to $OD2^{256}$. How many RC#2 matrices shall now move toward each of the 6 sides of each of the 6 RC#1 matrices? I've yet to make up my mind on that issue. On the one hand, a familiar progression inclines me to resort to successive squares of the powers of 2; on the other hand, another familiar progression inclines me to resort to successive squares of the powers of 6. So, I say this: In RC#2, a minimum of 2 and maximum of 6 RC#2 matrices shall move toward each of the 6 sides of each of the 6 RC#1 matrices. That, of course, means a minimum of 12 (*i.e.:* 2x6) and maximum of 36 (*i.e.:* 6x6) RC#2 matrices shall be in potency to each of the 6 RC#1 matrices. In turn, that means a minimum of 72 and a maximum of 216 RC#2 matrices. In turn, that means this: In the great eight-fold acceleration, RC#2 is created as 72 to 216 instances of the forms native to $OD2^{128}$ accelerate to somewhere between $OD2^{254}$ and $OD2^{255}-1$ and, thereby, produce 72 to 216 RC#2 matrices each of which can be *figuratively* described as a nest of roughly 10^{76} concentric boxes/cubes in which the smallest is roughly 10^{-47} cm. LB&D and the largest anywhere from roughly 4.5 (at $OD2^{254}+1$) to 9 (at $OD2^{255}-1$) trillion light years LB&D.

To express figuratively what RC#2 involves, turn back to the prior page's drawing, and focus on the 2 squares marked RC#1 and RC#2. While remembering the squares represent cubes, try to imagine this: 12 to 36 instances of RC#2 start at the center of RC#1 and then move toward the 6 sides of RC#1 by splitting into groups of 2 to 6, each of which moves toward a different side. Finally, imagine that process repeated in each of the other 5 RC#1 matrices not shown in the drawing.

How does this locomotion take place? Remember that the RC#1 *square* represents what is, figuratively speaking, a secondary *cube* composed of roughly 10^{232} primary cubes each roughly 10^{-47} cm. LB&D. *Figuratively* speaking, each of these primary cubes is a place in space which each RC#2 matrix's carrying generator can occupy. *Really* speaking, each of these primary cubes is a particular basic act which a carrying generator can perform, and the sequence in which a carrying generator must perform those acts is denoted by the position of each primary cube within the secondary cube to which it belongs. In short, each RC#2 matrix's locomotion from the center to some side of an RC#1 matrix taxes place as the carrying generator of the RC#2 matrix performs—in the required sequence—the acts made available to it by the RC#1 matrix.

In page 39's drawing, the one instance of RC#2 is shown as having gone all the way from the center to the A' side of the one instance of RC#1. In reality, it's not likely any instance of RC#2 has gone that far yet. I drew it that way only because of the limits imposed upon me by the size of these pages.

RC#3: How many RC#3 matrices shall move toward each of the 6 sides of each of the RC#2 matrices? One familiar progression tells me 2^2, and another tells me 6^2. So, I say this: A minimum of 4 and a maximum of 36 RC#3 matrices move toward each of the 6 sides of each of the 72 to 216 RC#2 matrices. That, of course, means a minimum of 24 (*i.e.:* 4x6) and maximum of 216 (*i.e.:* 6^2x6) RC#3 matrices shall be in potency to each of the 72 to 216 RC#2 matrices. In turn, that means a minimum of 1,728 and a maximum of 46,656 RC#3 matrices. In turn, that means this: In the great eight-fold acceleration, RC#3 is created as 1,728 to 46,656 instances of the forms native to $OD2^{128}$ accelerate to somewhere between $OD2^{252}$ and $OD2^{254}$-1 and, thereby, produce 1,728 to 46,656 RC#3 matrices each of which can be *figuratively* described as a nest of roughly 10^{76} concentric boxes/cubes in which the smallest is roughly 10^{-47} cm. LB&D and the largest anywhere from roughly 1.125 (at $OD2^{252}$+1) to 4.5 (at $OD2^{254}$-1) trillion light years LB&D.

To express figuratively what RC#3 involves, turn back to the drawing on page 39, and focus on the 2 squares marked RC#2 and RC#3. Need I say more that?

RC#4: How many RC#4 matrices shall move toward each of the 6 sides of each of the RC#3 matrices? One familiar progression tells me 2^4, and another tells me 6^4. So, I say this: A minimum of 16 and a maximum of 1,296 RC#4 matrices move toward each of the 6 sides of each of the 1,728 to 46,656 RC#3 matrices. That, of course, means a minimum of 96 (*i.e.:* 2^4x6) and maximum of 233,280 (*i.e.:* 6^4x6) RC#4 matrices shall be in potency to each of the 1,728 to 46,656 RC#3 matrices. In turn, that means a minimum of 165,888 and a maximum of 104,195,911,680 RC#4 matrices. In turn, that means this: In the great eight-fold acceleration, RC#4 is created as 165,888 to 104,195,911,680 instances of the forms native to $OD2^{128}$ accelerate to somewhere between $OD2^{248}$ and $OD2^{252}$-1 and, thereby, produce 165,888 to 104,195,911,680 RC#4 matrices each of which can be *figuratively* described as a nest of roughly 10^{74} concentric boxes/cubes in which the smallest is roughly 10^{-47} cm. LB&D and the largest anywhere from roughly 70 billion (at $OD2^{248}$+1) to 1.125 trillion (at $OD2^{254}$-1) light years LB&D.

To express figuratively what RC#4 involves, turn back to the drawing on page 39, and focus on the 2 squares marked RC#3 and RC#4. Here again, I assume I need say no more than that.

TOP HALF OF THE CHART

RC#	$x2^{256}$	OD RANGE	LB&D RANGE	# OF RC MATRICES BY 2^n
1	2^{-1}	2^{255}~2^{256}-1	$9x10^{12}$~$18x10^{12}$ ly	$1x6x1=6$
2	2^{-2}	2^{254}~2^{255}-1	$4.5x10^{12}$~9.10^{12} ly	$2x6x6=72$
3	2^{-4}	2^{252}~2^{254}-1	$1.125x10^{12}$~$4.5x10^{12}$ ly	2^2x6x72=1,728
4	2^{-8}	2^{248}~2^{252}-1	$7x10^{10}$~$1.125x10^{12}$ ly	2^4x6x1,728=165,888
5	2^{-16}	2^{240}~2^{248}-1	$2.75x10^8$~$7x10^{10}$ ly	2^8x6x165,888=254,797,824
6	2^{-32}	2^{224}~2^{240}-1	4,000~274,545,000 ly	2^{16}x6x($2.4x10^8$)=$1.00187x10^{14}$
7	2^{-64}	2^{192}~2^{224}-1	5.7 million mi~4,000 ly	2^{32}x6x10^{14}=$2.58x10^{24}$
8	2^{-128}	2^{128}~2^{192}-1	$5x10^{-8}$ cm~$5.7x10^6$ mi	2^{64}x6x($2.58x10^{24}$)=$2.86x10^{44}$

BOTTOM HALF OF THE CHART

RC#	# OF RC MATRICES BY 6^n
1	$1 \times 6 \times 1 = 6$
2	$6 \times 6 \times 6 = 6^3 = 216$
3	$6^2 \times 6 \times 6^3 = 6^6 = 46,656$
4	$6^4 \times 6 \times 6^6 = 6^{11} = 362,797,056$
5	$6^8 \times 6 \times 6^{11} = 6^{20} = 3,656,158,440,062,976 = 3.65615844 \times 10^{15}$
6	$6^{16} \times 6 \times 6^{20} = 6^{37} = 61,886,548,790,943,213,277,031,694,336 = 6.188455 \times 10^{28}$
7	$6^{32} \times 6 \times 6^{37} = 6^{70} = 2.955205 \times 10^{54}$
8	$6^{64} \times 6 \times 6^{70} = 6^{135} = 1.123101 \times 10^{103}$

CHART #1: STRUCTURE OF THE EIGHT REVERSE CATEGORIES.

I dare to imagine that the above chart gives all the information one needs regarding RC#5, 6, 7, and 8. On that basis, I shall decline to give those last 4 reverse categories the same treatment I gave to the first 4.

Line 8 in the bottom half of the above chart presents us with a problem. It says RC#8 can consist of 10^{103} cosmic super matrices. There can be no such thing as 10^{103} cosmic super matrices each headed by a leading form native to $OD2^{128}$. For, Esoptrics tells us that the number of such forms is 6^{128}, and $6^{128} = 4 \times 10^{99}$. Fortunately, Esoptrics easily provides a solution to the problem. Forms native to $OD2^{64}$ are also categorical forms and can be accelerated 2^{128} times to $OD2^{192}$. They, then, can serve as the leading forms at the head of the 10^{103} RC#8 matrices. Are there enough of them? Esoptrics tells us the number of them is 6^{192} which is to say 2.5×10^{149}. Yes, there's more than enough of them.

What is the significance of the 8 reverse categories? RC#1 first divides the universe into 6 sectors each able to serve as the center of, and anchor for, 12 to 36 of the 72 to 216 major super-clusters of galaxy clusters which comprise RC#2. Each of the 72 to 216 members of RC#2 then serve as the center of, and anchor for, 24 to 216 of the 1,728 to 46,656 minor super-clusters of galaxy clusters which comprise RC#3. Each of the 1,728 to 46,656 members of RC#3 then serve as the center of, and anchor for, 96 to 7,776 of the 165,888 to 362,797,056 clusters of galaxy clusters which comprise RC#4. Each of the 165,888 to 362,797,056 members of RC#4 then serve as the center of, and anchor for, 256 to 10,077,696 of the 254,797,824 to 3,656,158,440,062,976 galaxy clusters which comprise RC#5. Each of the 254,797, 824 to 3,656,158,440,062,976 members of RC#5 then serve as the center of, and anchor for, 393,216 to 16,926,659,444,736 of the 10^{14} to 6.2×10^{28} galaxies which comprise RC#6.[1] Each of the 10^{14} to 6×10^{28} members of RC#6 then serve as the center

---

[1] For Esoptrics, then, the *entire* universe contains as many as 6.2×10^{28} (line 6 of the lower half of chart #1) galaxies. For Esoptrics, though, the *entire* universe means 18 trillion light years in "diameter". Turn now to how many galaxies would be included in a single RC#5 matrix with a "diameter" of anywhere from .275 to 70 billion light years. Since there are as many as 3.656×10^{15} RC#5 matrices (line 5 of the lower half of chart #1), we must divide the 6.2×10^{28} by 3.656×10^{15}. Do that, and the re-

of, and anchor for, (2.6×10^{10}) to (8×10^{24}) of the (2.58×10^{24}) to (3×10^{54}) major star systems (*i.e.:* stars greater than 5.7 million miles in diameter) which comprise RC#7. RC#8 then provides RC#1 thru7 with atoms, asteroids, planets, stars and other bodies less than 5.7 million miles in diameter.[1]

In Esoptrics, $OD2^{128}$ is the "mirror threshold". As a result of the great eight-fold acceleration, we have 8 mobile categories *below* that threshold and 8 mobile reverse categories *above* it. Below are: OD1, OD2~3, $OD2^2$~15, $OD2^4$~255, $OD2^8$~ 2^{16}-1, $OD2^{16}$~2^{32}-1, $OD2^{32}$~2^{64}-1, and $OD2^{64}$~2^{128}-1. Above are: $OD2^{128}$~$OD2^{192}$-1, $OD2^{192}$~2^{224}-1, $OD2^{224}$~2^{240}-1, $OD2^{240}$~2^{248}-1, $OD2^{248}$~2^{252}-1, 2^{252}~ 2^{254}-1, $OD2^{254}$~ 2^{255}-1, and $OD2^{255}$~2^{256}-1 (#2^{256} is not mobile.). In today's Physics, there is an Eight-Fold Way. Wikipedia calls it the name given by American physicist Murray Gell-Mann to a theory which organizes subatomic baryons and mesons into octets. Is there some connection between that eightfold arrangement and the eightfold arrangement Esoptrics finds in the universe? I suspect there is, but have not the foggiest notion what that connection might be.

A-10: ESOPTRICS ON EXPANDING SPACE:

In its Internet article **Universe**, Wikipedia gives the age of the universe as 13.7 billion years plus or minus 120 million. How, then, does it manage to have a diameter of 93 billion light years? Shouldn't its *radius* be limited to how far light can travel in 13.7 billion years and its *diameter* thus be limited to twice that—namely: 27.4 billion years? It's a question which bugs many people.

The explanation commonly invoked is that of a balloon in its deflated versus its inflated state. Leaving the balloon deflated, we place 2 *living* insects on its surface. As long as we leave the balloon deflated, the speed with which the 2 insects move away from one another is determined solely by how fast they can crawl, and

---

sult is 1.7×10^{13} galaxies. Esoptrics thus says this: Within the confines of a single RC#5 matrix allowing it to expand to 70 billion light years, the universe contains 17 trillion galaxies. In its Internet article entitled **Observable Universe**, Wikipedia tells us the observable universe is 93 billion light years in diameter, but also mentions 78 billion as the diameter suggested by the cosmic microwave background radiation. The universe, it says, is "organized in more than 80 billion galaxies". Eight lines lower, it says: "There are possibly 80 billion galaxies in the Universe". Other Internet sources give a figure of 100 billion galaxies and admit the number may grow in years to come. That, I suggest, is not terribly far out of line with what Esoptrics says. Here, then, it would seem that Esoptrics—thru one man's pursuit of a purely introspective reasoning's analysis of the mind's inner world—leads to a conclusion somewhat in line with what Science manages to discover only by means of millions of scientists spending many billions of dollars upon efforts to examine the world outside of the human mind by means of microscopes, telescopes, and the like.

[1] Line 8 in the lower half of chart #1 suggests to me that the total number of atoms in the universe should be roughly 10^{103}. How many atoms, then, would be found in a single RC#5 matrix? To answer, we merely divide 10^{103} by the number of RC#5 matrices, which is to say 3.656×10^{15} according to line 5 in the lower half of chart #1. Do that, and the answer is 2.7×10^{86}. In its Internet article **Observable Universe**, Wikipedia tells us that there are approximately: "3×10^{79} hydrogen atoms in the observable Universe. But this is definitely a lower limit calculation, and it ignores many possible atom sources." Three lines higher, it gives a figure of 8×10^{79}. Here again, then, is not Esoptrics rather close to the mark?

that is limited to the speed of light. Next we glue 2 *dead* insects an inch apart on the balloon's surface. Being dead, there's no way they can move away from one another on their own; and yet, they shall indeed move apart from one another, if we inflate the balloon, and the speed at which they move apart is determined by how fast we inflate the balloon. They thus move apart because the *space* between them is expanding as that space curves. And why does it curve and expand? It does so, because it's going from a flat to a spherical shape. How can it do that and still be 3-dimensional? It's easy to imagine, if the balloon is doughnut shaped.

Here, then, the question is this: Can space expand faster than light can travel? To put it another way: Can space expand so fast, the 2 insects—though unable to move at all on their own—move apart at many times the speed of light? Modern Physics insists it can. How, now, does that relate to Esoptrics?

A'

1	2	3	4	5	6	7	8
9	10	11	12	13	14	15	16
17	18	19	20	21	22	23	24
25	26	27	28_0	29	30	31	32
33	34	35	36	37	38	39	40
41	42	43	44	45	46	47	48
49	50	51	52	53	54	55	56
57	58	59	60	61	62	63	64

A $_0$ B

C

DRAWING #6(REPEATED): ACTS AVAILABLE TO ANY GENERATOR RECEIVING ITS 3 SETS OF POWER/ANTI-POWER THRU A FORM CURRENTLY AT OD4

For Esoptrics, generators—as long as they are <u>not</u> locked inside the confines of an atom (***i.e.:*** as long as their piggyback form's OD is not less than 2^{128})—can never move faster than the speed of light; but, space can indeed expand at many times the speed of light. How is that possible? Basically, it's possible because, for Esoptrics, space is the actuality of a form. What has that got to do with it?

Generators, if you remember, can change from act to act in only one way. Using a familiar drawing repeated on the prior page, we can explain it so: If a given generator is occupying square 19 (*i.e.:* if it's performing the act figuratively described by square 19), the only way it can change internally is to move either to square 18 or square 20 or square 11 or square 27. Yes, the squares are actually cubes, and there are cubes directly above and below 19 to which 19 can "move". Pardon me, if I bow to the limits of my drawing skills and decline to do more than *mention* those other 2 cubes. As the given generator "moves from square to square" (*i.e.:* changes from act to act), it is subject to 2 limits: (1) It must follow the logical sequence figuratively described by the location of the smaller squares/cubes within the overall square/cube; and (2) it cannot—if outside the atom—change more frequently than 1 change per 2^{128}K (*i.e.:* 3.4×10^{38} alphakronons), which is to say the speed of light as defined by Esoptrics (*i.e.:* $1\Phi/2^{128}$K).

Forms, however, have 2 different ways to change from act to act. For, they can either remain at some given OD, or they can change to a higher or lower OD. Let's spell out those two options in that order.

While remaining at some given OD, forms expand and contract at regular intervals in the terms of alphakronons. How frequently they change is determined by a combination of their native OD and their current OD. Regardless of how frequently they expand or contract, when they do expand or contract, the change is *instantaneous*. For example, consider a form which is native to and currently at OD2^{128}. For 2^{256}K, its LB&D is $2^{129}\Phi$ (*i.e.:* .00 000 005 cm.). Then, suddenly, it is $2^{130}\Phi$ (*i.e.:* .00 000 001 cm.) for the next 2^{256}K. That's not much of an instantaneous change. Consider though a form which is native to, and currently at OD2^{192}. For 2^{384}K, its LB&D is 5.7 million miles, and then, suddenly and instantaneously, its LB&D is 11.4 million miles for the next 2^{384}K (*i.e.:* 3.9×10^{115} alphakronons).

What effect does that expansion and contraction have upon the generators in potency to that form? Does it cause the two dead insects to move apart at greater than the speed of light? Esoptrics says it does not. What, then, does it do?

Let's answer by again using the prior page's drawing. Imagine, then, a generator performing the act represented by square/cube 1. Doing that, it's at the outermost reaches of the overall square/cube represented by the drawing, and its distance from the center of that overall square/cube is 7Φ (assuming generators cannot move diagonally). As the generator is at square/cube 1, the overall square/cube doubles its LB&D. Shall the generator suddenly double its distance from the center of the overall square/cube? Modern Physics may answer yes; but, Esoptrics answers no. Esoptrics says, instead, that the generator must move on its own toward the new outer boundary, and, as it does so, it can move no faster than $1\Phi/2^{128}$K. Esoptrics, then, contradicts modern Physics and says expanding space does not push any of its occupants apart from one another, but does gives some of its occupants *room* to move further apart from one another.

Suppose, as the generator reaches the new outer limits of the overall square/cube in drawing #6, the overall square/cube suddenly reverts to its prior LB&D? What shall become of the generator? Moving 1Φ per change (and now using the *potential* rather than the *actual* primary cubes of the form/secondary-cube to which it is in potency), it shall plunge back toward the contracted boundary of the form to which it is in potency. Its velocity perhaps increases as a result of the plunge; but, that velocity shall still not exceed the speed of light.[1]

That, then is how forms expand and contract while remaining at a given OD. What happens to a form's LB&D when it accelerates? Imagine a form is native to $OD2^{128}$, and it accelerates to $OD2^{256}$-1. As has been said earlier, it cannot, in doing so, skip any of the intermediary ontological distances. Thus, to go from $OD2^{128}$ to the projected OD, it must first perform an act at $OD2^{128}$+1, then an act at $OD2^{128}$+2, then $OD2^{128}$+3, and so forth. Every *generator* must tarry at least 2^{128}K in every one of its acts. How long must an *accelerating form* tarry in each of its acts on its way to its destination OD? The answer is *one* alphakronon per OD. In short, every accelerating form increases its LB&D at 2^{128} times the speed of light. Again, though, the accelerating form (*i.e.:* expanding space envelope) does not cause any of its occupants to move apart from one another faster than the speed of light. Yes, it gives some of its occupants an expanded area in which to move apart on their own at no more than the speed of light; but, it does not by any means expand the length of the space between them. In short, it gives them more room in which to roam, but doesn't, on its own, expand the space *between* them.

A-11: ESOPTRICS ON ANTI-GRAVITY:

Esoptrics' doctrine on acceleration also explains what anti-gravity is. How so?! Suppose two forms are each native to the same *categorical* OD, such as 2^{128}. Next, suppose they are both accelerated to distances which are both within the confines of the same reverse *category*, such as RC#6 running from $OD2^{224}$ to $OD2^{240}$–1. If so, equally accelerated like shall repel equally accelerated like, and the result will be two galaxies moving away from one another as the super matrices at their centers repel one another.

By the same token, suppose two forms are each native to the same *genus*, such as 2^{127}. Next, suppose they are both accelerated to distances which are both within the boundaries of the same *genus*, such as the genus comprised of $OD2^{192}$ thru $OD2^{193}$–1. If so, equally accelerated like shall repel equally accelerated like, and

---

[1] In the end, this relationship between generators and expanding and contracting forms is Esoptrics' way of explaining the elliptical orbits of the planets. Earth does not orbit the Sun. At Earth's center is a super matrix whose carrying generator is in potency to, and at the outermost boundary of, a form centered in the Sun and rotating around its center in the Sun. In a manner of speaking, Earth's carrying generator is the one occupying square 1 in drawing 6; and so, as the overall square rotates around its center in the Sun, it causes Earth, too, to orbit the Sun. When the form abruptly expands, Earth's carrying generator—as it heads for its form's newly *increased* outer limits—necessarily moves *away* from the Sun, and, when the form contracts, Earth's carrying generator—as it heads for its form's newly *reduced* outer limits—moves *toward* the Sun. For Esoptrics, when the ancients said angels move the planets, they were closer to the truth than is modern Science, since, for Esoptrics, the forms are indeed a kind of angelic reality.

one of the results will be a planet circling a star, because pressed to the star's *central* super matrix is a smaller one which is the twin of the planet's *central* super matrix. Yes, a super matrix with a *categorical* leading form is very different from a super matrix with a *generic* leading form.

A-12: ESOPTRICS ON 127 ELEMENTS:

If forms native to $OD2^{128}$ are accelerated much beyond $OD2^{128}$, they spawn hydrogen atoms. If forms native to $OD2^{127}$ are accelerated much beyond $OD2^{128}$, they spawn helium atoms. If forms native to $OD2^{126}$ are accelerated much beyond $OD2^{128}$, they spawn lithium atoms, etc..

Esoptrics thus allows for up to 127 elements. Forms exactly at $OD2^{128}$ and not part of a super matrix are a very important kind of radiation.

A-13: ESOPTRICS ON VELOCITY:

Especially above 2^{127}, if a matrix, super matrix, or cosmic super matrix (except for 2^{256} since it has no form to which it can be in potency) is *exactly* at a generic number of steps from zero actuality, its velocity is generally light speed (*i.e.:* $1\Phi/2^{128}K$). If not, its velocity = $S_{AG}(C/N_G)$, where C = light speed, N_G = current generic distance, and S_{AG} = the number of steps above that generic distance.

For example, if a cosmic super matrix is at $OD2^{255}$, it is moving at light speed; but, if it accelerates one more step from zero actuality to $OD2^{255}+1$, then its velocity drops to light speed divided by 2^{255} then times 1. If it accelerates to $OD2^{255}+2$, then its velocity is light speed divided by 2^{255} then times 2.

Esoptrics thus has as many as 128 kinds of light speed radiation, which is to say 1 per power of 2 above 2^{127} except for 2^{256}. Within each of those 128 kinds, an astronomical number of frequencies are possible.

A-14: ESOPTRICS ON DARK MATTER & 4 DIMENSIONAL SPACE:

Esoptrics also has an explanation for what dark matter is. Its explanation has to do with the fact that, for Esoptrics, space—even without taking time into consideration—is 4 rather than 3 dimensional. How, then, is space 4 dimensional?

To make the answer as simple as possible, let's use the drawing on the next page. Combine imagination with one's eyes, and one square/cube figuratively represents the form of the universe, while a second square/cube figuratively represents a first reverse category, while a third stands for an RC#2 and a fourth for an RC#3.

At first glance, the drawing seems to be saying that each of the matrices RC#1, 2, and 3 is moving toward the outer limits of the form of the universe ("Omega256" for short). That, though is not what the drawing is saying; instead, it's saying only RC#1 is *truly* moving to, and never beyond, the limits of Omega256. The matrices RC#2 and 3 are only *coincidentally* moving toward Omega256's limits. *Actually*, RC#2 is moving toward the outer limits of RC#1. That's a limit which may eventually put the center of RC#2 as much as 9 trillion light years (and more on the

cubes, the largest of which is anywhere from 1.125 to 4.5 trillion light years in LB&D, and the smallest of which is 4Φ in LB&D.

How does such nesting produce a kind of space which is 4 dimensional? It does so by producing a kind of space which is stratified in the presence of massive bodies such as galaxies, stars, planets, and what have you. What does that mean?

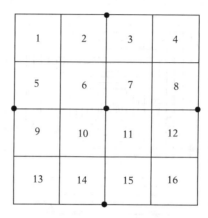

DRAWING #11: A FORM AT OD2.

In your imagination, take the above drawing, and impose it upon drawing #6 on page 44. Put them together in such a way that square 1 above = square 19 in the other; square 2 above = square 20 in the other; 3 above = 21 in the other; 4 = 22; 5 = 27; 6 = 28; 7 = 27; 8 = 30; 9 = 35; 10 = 36; 11 = 37; 12 = 38; 13 = 43; 14 = 44; 15 = 45; and 16 = 46. That's a nest only 2 boxes/cubes deep. That's as far as this book's physical limits allow me to go. Well, I probably could throw in a figurative description of a form at OD3; but, I'll plead exhaustion and drop the opportunity.

To "carnal" minds able to think only in the terms of "Geometry speak" and what's *physically* separate in space, squares 1 thru 16 above have ceased to exist, and: square 19 is square 19 and only square 19; square 20 is square 20 and only square 20; square 21 is square 21 and only square 21; etc. all the way to square 46. In reality, though, it makes no difference how totally squares 1 and 19 are *physically* one and the same; they are *logically* outside of one another in the order of sequence observed by the various ontological distances. Square #1 is "where" it is relative to OD2, and square 19 is "where" it is relative to OD4, and never the twain shall become one and the same *logically*. The same is true of 2 and 20, 3 and 21, 4 and 22, etc. all the way to 16 and 46.

But, what's the point of it all? Imagine 2 generators are moving away from the center of the nest. One is in potency to, and is using, the form at OD2. The other is in potency to, and is using, the form at OD4. The former can go no further from the center of the nest than squares 1, 2, 3, 4, 5, 8, 9, 12, 13, 14, 15 or 16. The latter can go no further from the center of the nest than squares 1 thru 8, 9, 16, 17, 24, 25, 32, 33, 40, 41, 48, 49, 56, or 57 thru 64. To "carnal" minds able to think only in the terms of "Geometry speak" and what's *physically* separate in space, it will seem the two generators are moving thru the same one area of the same one, good old-fashioned, homogeneous, 3 dimensional space.[1] They are not. They are moving thru 2 very different layers within the same one, 4 dimensional space envelope, which is to say within the same one, 4 dimensional super matrix created by a form which has accelerated beyond its native OD.

Have you followed that? If so, then, hopefully, you can begin to understand what dark matter is for Esoptrics. What it boils down to is this: Like everything else moving thru space, light is moving thru a *four* dimensional medium and, therefore, must move thru that medium at one or more of that medium's many *levels*. As a result, if we are to receive the light from some particular source, it is not enough for that source's light to pass thru our area of the universe. In addition, it must move thru our area at one or more of the levels in 4 dimensional space to which our area of the universe has access.

For example, light radiating from sources within any of the 6 instances of RC#1 generally travel thru their RC using a form well above $OD2^{255}$. If the center of some instance of RC#2 is logically far enough away from the center of every instance of RC#1, almost the half of RC#2's occupants shall be in an area where no OD higher than $2^{255}-1$ is logically present; and so, none of those occupants can possibly receive light traveling at an OD well above $OD2^{255}$.

I strongly suspect we are in an area of the universe in which, logically, our instance of RC#8 is still wholly within the limits of an instance of RC#5 whose center is so far from every instance of RC#4 that, for us, the highest OD logically present to us is $2^{248}-1$ (See line 5 of the upper half of chart #1 on page 42.). If so, that would explain why we receive none of the light from any instance of either RC#4 or RC#3 or RC#2 or RC#1. For, in each of those cases, light is traveling at an OD far above 2^{248}—namely: $2^{252}-1$, $2^{254}-1$, $2^{255}-1$, and $2^{256}-1$ respectively.

To see a figurative explanation of that, examine the drawing on the next page for a bit. Do that adequately, and you should notice the following:

An instance of RC#2 has gone so close to the outer limits of an instance of RC#1, that many of the largest boxes/cubes in RC#1's nest of concentric forms logically fail to extend to more than about a quarter of RC#2. Since light coming from sources in RC#1 travels thru the universe at a level well above $OD2^{255}$, most of the occupants of RC#2 shall never see any of that light.

Likewise, an instance of RC#3 has gone so close to the outer limits of an instance of RC#2, that many of the largest boxes/cubes in RC#2's nest of concentric forms logically fail to extend to more than about a quarter of RC#3. Since light coming from sources in RC#2 travels thru the universe at a level well above $OD2^{254}$, most of the occupants of RC#3 shall never see any of that light.

Likewise, an instance of RC#4 has gone so close to the outer limits of an instance of RC#3, that many of the largest boxes/cubes in RC#3's nest of concentric forms logically fail to extend to more than about a quarter of RC#4. Since light coming from sources in RC#3 travels thru the universe at a level well above $OD2^{252}$, most of the occupants of RC#4 shall never see any of that light.

Finally, an instance of RC#5 has gone so close to the outer limits of an instance of RC#4, that many of the largest boxes/cubes in RC#4's nest of concentric

---

[1] And they will puzzle greatly over why one goes further from the center than the other.

forms logically fail to extend to more than about a quarter of RC#5. Since light coming from sources in RC#4 travels thru the universe at a level well above $OD2^{248}$, most of the occupants of RC#5 shall never see any of that light.

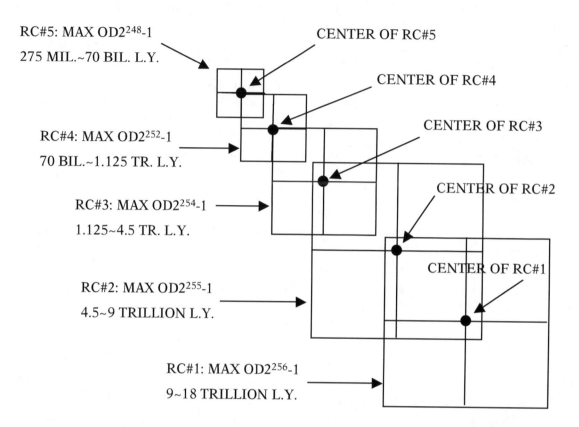

RC#5: MAX $OD2^{248}$-1

275 MIL.~70 BIL. L.Y.

CENTER OF RC#5

CENTER OF RC#4

CENTER OF RC#3

RC#4: MAX $OD2^{252}$-1

70 BIL.~1.125 TR. L.Y.

RC#3: MAX $OD2^{254}$-1

1.125~4.5 TR. L.Y.

CENTER OF RC#2

CENTER OF RC#1

RC#2: MAX $OD2^{255}$-1

4.5~9 TRILLION L.Y.

RC#1: MAX $OD2^{256}$-1

9~18 TRILLION L.Y.

DRAWING #12: LIGHT PATHS BETWEEN REVERSE CATEGORIES.[1]

On the other hand, an instance of RC#5 may still have well within its limits every instance of RC#6 in potency to it, every instance of RC#7 in potency to those instances of RC#6, and every instance of RC#8 in potency to those instances of RC#7. That's very possible because the drop from RC#5 to RC#6 is so much more dramatic than the drop from RC#4 to RC#5. In other words: Though every instance of RC#4 is only 256 (*i.e.:* 2^8) times as long, wide, and deep as every instance of RC#5, every instance of RC#5 is 65,536 (*i.e.:* 2^{16}) times as long, wide, and deep as every instance of RC#6, 4.3 billion (*i.e.:* 2^{32}) times as long, wide, and deep as every instance of RC#7, and 18 quintillion (*i.e.:* 2^{64}) times as long, wide, and deep as every instance of RC#8. With all those instances of RC#8, 7, and 6 well within the confines of an instance of RC#5, every source of light within the confines of that

---

[1] I risk unduly complicating things by doing so; but, I must point out that each of the above 5 squares/cubes is rotating around its own center—RC#5 faster than RC#4, RC#4 faster than RC#3, RC#3 faster than RC#2, and RC#2 faster than RC#1. As you can perhaps see for yourself, that must eventually produce a telescoping action. Fortunately, each of the above squares/cubes rotates so slowly, the outermost segment of the "telescope" won't wind up inside the innermost segment of the "telescope" for trillions of years by our calculation of time.

instance of RC#5 has the *potential* to cast its light upon all the other occupants of RC#5.

I say has the *potential*, because there's still the problem that light running around in RC#5 can do so at 2^{248}-1 different levels within RC#5's 4 dimensional space envelope. As a result, for Esoptrics, there are 2 kinds of dark matter: (1) what's dark because its light emissions travel thru space at an OD higher than any OD in our part of the universe, and (2) what's dark because its light emissions travel thru space at an OD which is normally in our part of the universe but which, for some reason, is *locally* missing or obstructed. In the latter case, dark matter is dark for us due to some kind of OD gap in our vicinity. That, of course, implies that what is dark matter for us on Earth may not be dark matter for other observers elsewhere in our solar system or galaxy.

In a *physically* extended universe, all that matters is whether or not a given source of light (or any other kind of radiation) is within the universe. In a *logically* extended universe, what matters is whether or not there is a *logical* path between a given observer and a given source of light. If there is no *logical* path, it shall seem to the observer that the given source is giving off no light and, therefore, is dark matter. It is possible, of course, that, though there may be no logical path between the observer and the source's *light*, there may be a logical path between the observer and some other kind of radiation from the source. Individuals much smarter than myself shall have to figure out those logical paths.

A-15: RECTILINEAR VS. CURVILINEAR LOCOMOTION:

If sharp, my reader shall have noticed there's been no mention of *curvilinear* motion so far. So far, all mention of "locomotion" has been about generators, and they move only in a *rectilinear* fashion. Whence, then, comes curvilinear motion?

For Esoptrics, all curvilinear motion comes about because the forms, in most cases, rotate on their axes as a result of being "in potency" to two extra-cosmic influences which Esoptrics calls "heavenly sextets" and "heavenly benchmarks". Let's deal with them in that order.

Every form has its own heavenly sextet, and the logical distinction between one such sextet and another is that no two of them are associated with the same form. The forms are thus the means by which the heavenly sextets attain logical distinction and, consequently, their own unique individuality.

The members of each such sextet, I designate by the symbols: (1) A, (2) A' (*i.e.:* A prime), (3) B, (4) B' (*i.e.:* B prime), (5) C, and (6) C' (*i.e.:* C prime). Each of these 6 is a reference point in infinity, and each has a unique effect upon the form with which it is associated. Mainly, each gives its associated form an absolute *logical* orientation which we can *figuratively* and *physically* express by recourse to the imagery of a cube. That has 12 logical results: (1) The A side of the form/cube is forever such and is directly opposite the B side; (2) the B side of the form/cube is forever such and is directly opposite the A side; (3) the C side of the form/cube is forever such and is directly opposite the A' side; (4) the A' side of the form/cube is forever such and is directly opposite the C side; (5) the B' side of the form/cube is forever such and is directly opposite the C' side; (6) the C' side of the form/cube is

ample, if a form native to $OD2^{128}$ is still at that OD, then its LB&D (*i.e.:* "diameter" for "Geometry speak") is $2^{129}\Phi$; and so: (1) the duration of each state in the tertiary plane = (.5L)K, where L = the LB&D of the associated form in Φ; and (2) the duration of each state in the secondary plane = $(\sqrt{.5L})K$ and ±1 if the square root is not a whole number. On now to this document's main issue!

B: CONSCIOUSNESS:

As with so much else in the universe, every act of consciousness is a well-defined, primary act on the part of a matrix (*i.e.:* form), and each act has a definite duration and "size" relative to all other acts by all other matrices. Consciousness is thus a series of quanta. Which matrix, then, performs our acts of consciousness, and what is its "size" and duration (*i.e.:* rate of change relative to the maximum)?

Esoptrics says consciousness is a series of individual primary acts (and thus quanta) on the part of a form *native* to, and *currently* at, somewhere between $OD2^{155}$ and $OD2^{156}$. For what follows, it's not necessary for us to know either *exactly* where it is or *why* it's there.

Applying formula #1 on page 36, we calculate each cycle's first act—on the part of our #2^{155} form—is a cube somewhere between 2^{156} and $2^{157}\Phi$ LB&D. That's somewhere between 9.1343852×10^{46} and $1.82687704 \times 10^{47}\Phi$ LB&D. That, in turn, translates into somewhere between 6.71 and 13.42 cm. LB&D, which is to say 2.64 to 5.28 in. LB&D where 2.54 cm. = 1 in.. That fits quite nicely with the size of the human brain which Wikipedia, in its article **Human Brain**, gives as 1130 cubic cm. (*i.e.:* the cube of 10.4158 cm.) in women and 1260 cubic cm. (*i.e.:* the cube of 10.8008 cm.) in men

Applying formula #5 on page 36, we calculate every cycle of such a form would have a *minimum* duration of $2[(2^{155})^2] = 2(2^{310}) = 2^{311} = 4.1718496 \times 10^{93}$K and a *maximum* duration of $2[(2^{156})^2] = 2(2^{312}) = 2^{313} = 1.66873987 \times 10^{94}$K. As you can perhaps readily perceive, it would have that minimum if it were exactly at 2^{155} and that maximum if it were exactly at 2^{156}.

Esoptrics says that 1 of our seconds = $1.388543802 \times 10^{95}$K. Dividing that figure first by 4.1718496×10^{93} (*i.e.:* 2^{311}) and then by $1.66873987 \times 10^{94}$ (*i.e.:* 2^{313}), we calculate our form would, in one second, perform somewhere between 33.28 and 8.32 *cycles* at 2^{155} and 2^{156} respectively. Since each of those cycles involves 2 acts, our form would, in one second, perform somewhere between 66.56 *acts* at 2^{155} and 16.64 *acts* at 2^{156}, and each of those *acts* would last somewhere between 1/66.56 seconds (*i.e.:* .015024 sec.) at 2^{155} and 1/16.64 seconds (*i.e.:* .060096 sec.) at 2^{156}.

That's interesting to note for a very important reason: If I remember correctly, it's been well established by observation that we humans cannot detect the flashing nature of what flashes more than 20 times a second. For example, an old note of mine says that, on page 72 of vol. 23 of the 1967 edition of **Encyclopedia Britannica**, we're told that, when extremely bright flashes of light are played on the eye, the eye is totally unable to detect the flashes once the rate reaches approximately 20 per second. If that's correct then the form we're describing here would correlate nicely with that. For, we are saying there are somewhere between 16.64

and 66.56 acts of consciousness per second, and the rate of 19 to 20 acts of consciousness per second falls within that range. If you rightly understand how static a single act is, it follows that, if consciousness acts *less* than 20 times a second, then it won't be able to detect the flashing mode of a light flashing *more* than 20 times a second. I'll go into that more thoroughly on the next page.

Thus does Esoptrics' appeal to a form somewhere between 2^{155} and 2^{156} correlate nicely with what scientific observation says about the size of the human brain and the human inability to detect the flashing nature of light once it flashes 20 times a second or more. But, that correlation comes about only if Esoptrics' formulas for calculating size and absolute rate of change are correct when giving the form: (1) a size between 2^{156} and $2^{157}\Phi$, and (2) a change rate of between once every 4.1718496×10^{93}K and once every $1.66873986 \times 10^{94}$K. In turn, the correctness of those formulas is based upon these two hypotheses: (1) The minimum size = $7.34683969 \times 10^{-47}$cm., and (2) the minimum length of time = $7.201789375 \times 10^{-96}$ seconds, thus making the maximum rate of change $1.388543802 \times 10^{95}$ times per second. Consequently: (1) In fitting the size of the form at 2^{155} to 2^{156} to the size of the brain, Esoptrics strongly upholds its contention regarding the minimum unit of space, and (2) in explaining the flashing light syndrome by means of its formula #5, Esoptrics thus strongly upholds its contention regarding the minimum unit of time and maximum rate of change. That it could correlate two ways like that is no small matter.

So then, we have two time frames: one involving 66.56 acts per second and another involving 16.64 acts per second. Let us now calculate how far light travels in each of those time frames.

If we take the speed of light as 29,979,245,800 cm./sec. and divide it by 16.64, the result is 1,801,637,367.79 cm. for the form at 2^{156}. Divide by 66.56, and the result is 450,409,341.95 cm. for the form at 2^{155}. Those 2 results must then be doubled, since we receive light from opposite directions. 1,801,637,367.79 thus becomes 3,603,274,735.58 cm., and 2 x 450,409,341.95 = 900,818,683.9 cm..

So then, if the form of consciousness is exactly at 2^{155}, it's performing 66.56 acts per second each open to the *bi-directional* influx of light from an area 900,818,683.9 cm. in diameter. This light it must pack into a cube only 6.71 cm. LB&D. Obviously, we're going to have to do a lot of miniaturizing. Dividing 900,818,683.9 cm. by 6.71, we find the number of times such light must be reduced, and that number is 134,250,176.438 times. What's that multiple's square root? My calculator says it's 11,586.6378. Reduce 6.71 cm. by that square root. The result will be .00 057 9 cm.. That, I dare *suggest*, is about the size of the smallest object consciousness can detect per act. Next, reduce .00 057 9 cm. by that same 11,586.6378. The result will be .00 000 004 998 cm., and that puts us at the "size" of the hydrogen atom. That amounts to saying this: Take the atomic diameter (*i.e.:* the mirror threshold's diameter a/k/a 2^{128} steps from 0 actuality), multiply it by 11,586.6378, and that gives you the *minimum* diameter of what consciousness can sense per act; next multiply the atomic diameter by $11,586.6378^2$, and that gives you the *maximum* diameter of what consciousness can sense per act; next multiply the atomic diameter by $11,586.6378^4$, and that gives you the bi-directional distance light travels per act of consciousness.

Does that progression sound familiar? It should, because, in Esoptrics, the progression N, N^2, N^4, etc., is the one found among the categorical forms, and the categorical form #2^{128} rules the complete atom.

Suppose the form of consciousness is at 2^{156}. It would not be *exactly* there, since it would have to be at least 1 step short of that. Still, working with 2^{156} rather than something such as 2^{156}–1, is more than adequate for our purposes. At 2^{156}, it's performing 16.64 *acts* per second each open to the *bi-directional* influx of light from an area 3,603,274,735.58 cm. in diameter. This light it must pack into a cube 13.42 cm. LB&D. Here, too, we're going to have to do a lot of miniaturizing. Dividing 3,603,274,735.58 by 13.42, we find the number of times such light must be reduced, and that number is 268,500,352.87. What's that multiple's square root? My calculator says 16,385.98 (2^{14} = 16,384). Reduce 13.42 cm. by that square root. The result will be .00 081 9 cm.. That, I dare *suggest* is about the size of the smallest object consciousness can detect per act. Now, reduce .00 081 9 cm. by that same 16,385.98. The result will be .00 000 004 998 cm.. Once again we're at the "size" of the hydrogen atom. Once again, that amounts to saying this: Take the atomic diameter, multiply it by 16,385.98, and that gives you the *minimum* diameter of what consciousness can sense per act; next multiply the atomic diameter by 16,385.98^2, and that gives you the *maximum* diameter of what consciousness can sense per act; next multiply the atomic diameter by 16,385.98^4, and that gives you the bi-directional distance light travels per act of consciousness. Here again, it's the familiar progression of N, N^2, N^4.

Thus does Esoptrics take the 6 formulas given on page 36 and correlate them with: (1) the size of the human brain, (2) the maximum and minimum diameters of what consciousness can sense per act, (3) the fact consciousness cannot detect the *flashing* nature of light flashing more than 20 times per second, (4) the notion that the minimum unit of time is 7.201789375 x 10^{-96} seconds, (5) the notion that the minimum unit of space is 7.34683969 x 10^{-47} cm., (6) the diameter of the hydrogen atom, and (7) the speed of light. Who else has done as much?

C: AWARENESS OF TIME & CHANGE:

Just how static is every single act of consciousness? To answer, let's first compare the duration of our form's two act cycle (***viz:*** $2[(2^{155})^2]$ = 2^{311} = 4.1718496 x 10^{93}K) with the duration of the two act cycle of a form native to, and still at, 1 step from zero actuality. That latter duration is only 4K. In other words, for every cycle on the part of matrix #2^{155}, there are 10^{93} cycles on the part of matrix #1. From the viewpoint of matrix #1, every cycle on the part of matrix #2^{155} is roughly 10^{93} times longer in duration than what matrix #1 would call roughly a 20th of its version of a second; therefore, what matrix #2^{155} experiences as taking place in the flash of a 20th of a second, matrix #1 experiences as taking an astronomical number of millennia to run its course (***viz:*** by our reckoning, 10^{92} seconds = 10^{82} millennia).

But how can matrix #2^{155} experience a duration that astronomically lengthy and experience it as being quite brief? It's because, as long as matrix #2^{155} is performing the same one primary act, it undergoes no internal change whatsoever and, consequently, is aware of no kind of change whatsoever. With no awareness of any kind of change whatsoever, it has no clue whatsoever to the duration of its primary act; and so, its act seems, to it, wholly without duration. As you know, if you sleep soundly, 8 hours go by in a flash.

How, then, does matrix #2^{155} experience change? Like all matrices (*i.e.:* forms), matrix #2^{155} performs 2 kinds of primary acts: an expanded one followed by a contracted one. In its expanded act, matrix #2^{155} is aware of what's frozen and un-changing in the primary cubes its embracing; whereas, in its contracted act, it is aware of nothing (Thus, as far as consciousness is *aware*, it performs only one act per cycle.). While in its contracted act, an immense series of generators each act an astronomical number of times and, thereby, deposit units of force (tension) in each of the $(2^{156})^3$ (*i.e.:* $2^{468} = 7.62145 \times 10^{140}$) cubes which matrix #$2^{155}$ will experience when it performs its next expanded act. In experiencing these units of tension, ma-trix #2^{155} feels one or more of the different kinds of sensation. For *consciousness*, matter is neither more nor less than the units of tension generated by the genera-tors. And what a divide there thus is between intra-mental sense images and those extra-mental realities which are other than units of tension!

Performing that next expanded act, matrix #2^{155} is not observing what *is hap-pening* but, rather, the *residue* of what *happened* in a past which, from matrix #1's point of view, happened over the course of an astronomical number of millennia. That residue, of course, is spread out over primary cubes only *logically* outside of one another; but, all matrix #2^{155} can observe is that they are *outside of one another* and, not being able to observe the *manner* in which they are outside of one another, merely assumes they are *physically* outside of one another in *space*. Shades of Im-manuel Kant!

Still, how is there awareness of change? The brain, thru memory, somehow allows form #2^{155} to compare the content of its subsequent primary acts with the content of its previous ones. How might that be possible? I suspect it has to do with the difference in the "sizes" of the contracted vs. the expanded act. Since the latter is twice that of the former, it's possible that, in the latter, the form experiences the content of both its latest and its prior contracted act. Or maybe it's like this: Half of its 10^{140} primary cubes present it with the extra-mental world's effects, while the other half presents it with its intra-mental world's effects.

How does matrix #2^{155} live in and experience a world radically different from the world matrix #1 lives in and experiences? In other words, how can the latter be, not merely a part of the former but, rather, a subservient part of it? The answer is first this: Matrix #2^{155} lives in and experiences a world populated by other matrices each having a current distance from zero actuality somewhat close to its own. Ma-trix #2^{155} is thus a kind of strobe which causes the rapidly rotating fan blades to be-come stationary blades and, thereby, matrix #2^{155} experiences what's going on from the standpoint of matrices changing at or near its own rate of change.

D: THE FITFUL MICROSCOPIC WORLD VS. THE ORDERLY MACROSCOPIC ONE:

As for how matrix #1 is a subservient part of the world of matrix #2^{155}, bear in mind that every #1 matrix encompasses only 8 primary cubes; whereas, matrix

#2^{155} encompasses roughly 10^{140}. It's merely a case of the _macro_scopic exercising its influence over far more of the _micro_scopic than the microscopic itself can.

Science, these days, puzzles over how the cloudy and fitful world of microscopic particles gives birth to our clear and reasonably orderly macroscopic world. The problem here is that Science knows _virtually_ nothing about forms and, therefore, virtually nothing about the universe's _macro_scopic ultimate particles and how they connect to, and govern, the universe's _micro_scopic ones. Only Haasian cosmology really throws any light on that, and the light it throws is far from paltry.[1]

Yes, only Esoptrics details the _macro_scopic ultimates and how they connect to, and govern, the microscopic ones. For all of that, to this cosmology, both scientists and philosophers have—for 47 years now—turned an ear as deaf as any ear can be. Let them _continue_ to do so, and let's see what history one day says of them.

Such, then, is what Esoptrics says about consciousness. It is no small achievement, is it not? And from whom has it come? It has come from a high-school graduate with little or no knowledge of physics or higher math and who—abandoning Science's devotion to examining every one's _extra_-mental world—turned instead to Philosophy's (Mysticism's?) devotion to examining one's own _intra_-mental world, while also ceaselessly pounding on God's door and most anxiously begging for the gift of high proficiency in the use of contemplation rather than in the use of telescopes and microscopes.

> What a great practical joke nature will have played on us if all the thinking that has gone into discovering the ultimate laws of the universe turns out to reveal that one of the biggest clues was woven all along into the very fabric of thought itself.
> ——**DAVID H. FREEDMAN: _Quantum Consciousness_**, as found on page 98 of the June, 1994, edition of **_Discover_** magazine.

Behold, Mr. Freedman, sir! The great practical joke is upon us big time.

---

[1] Perhaps an even more significant problem is that Science knows _absolutely_ nothing about the Heavenly Sextets, the Heavenly Benchmarks, the ability of the former to "rotate roles", and the ability of the latter to form a hierarchical structure. The ability of the former to rotate roles is what produces atoms composed of sub-atomic particles with "color" and "charge", and the ability of the latter to form a hierarchical structure is what produces inorganic and organic molecules, crystals, plants, animals and human bodies. The human soul is then that instance of a Heavenly Benchmark which, by the grace of God, is created and placed at the very top of a hierarchical structure formed by a group of Heavenly Benchmarks, which, at their lowest level, govern the Heavenly Sextets controlling the movements of the ultimate building bricks of the universe (In Transubstantiation, Jesus' hierarchical structure supplants the usual one.). It is enough for now. Besides, no one ever listens to me anyhow. A learned nephew's explanation of that is: "Uncle Edward, no one can understand anything you write." History, if only in a distant future's Day of Judgment, shall give the final answer.

Document #3:

CIRCULAR LETTER STARTED MAY 11, 2007[1]

A: AN OPENING COMPLAINT:

The Philosophical Community
The Scientific Community &
The Entire Academic World

Greetings:

This rather lengthy letter of complaint is a 71 year old male's outcry against the 45 years of utter indifference which you have manifested toward my many ideas. I protest the gross injustice of your inaction toward my ideas. More importantly, I shall here set before the world, for all time to come, a bill of indictment which I hope shall prove my contention that your indifference to my ideas *is* in fact a gross injustice. This bill of indictment, I shall then send to hundreds if not thousands of sources with the intention of raising such a noise, that my complaint against you shall stand forever in the pages of history.

What evidences, then, do I present against you? I present my thinking's answers to some of the major questions which have long plagued the academic world, and let us then see if, at long last, some of you are able to recognize the importance of those answers. If not, then, hopefully, enough copies of this complaint shall survive as to insure that, someday, a future generation shall appreciate the importance of my ideas and, so doing, shall pronounce, against my contemporaries, the condemnation I say they so richly deserve.

---

[1] **NOTE OF SATURDAY, JULY 11, 2009:** Between this last and 2 earlier versions of it, some 1,750 copies of this letter went out in May, June, and July of 2007. I sent one copy each to the Philosophy and/or Physics departments of roughly 1,000 universities (Not all had both departments.). To that massive effort, there were only 4 responses. Two were quite terse and derogatory. A third was positive for a short while. A fourth was in the form of a friendly e-mail from a rather renowned professor serving as the chair of the Philosophy department of a major university in the state of New York. He was obviously trying to help me; but, when I read his e-mail, I realized he had totally misunderstood what I was trying to say. I e-mailed him back asking for permission to clarify my concepts for him. He gave me that permission, and, within the next few months, I sent him two letters. They follow this document as documents #4 and #5. No, he never replied to the letters. Much in this letter is redundant and poorly stated; but, I include it anyhow: (1) because of what it says about the EPR paradox, (2) because it lays the groundwork for the next 2 documents, and (3) because of section G.

B: ESOPTRICS ON THE EPR EFFECT:[1]

One of the scientific world's great questions has to do with an issue first raised, I'm told, by the joint effort of Albert Einstein, Boris Podolski, and Nathan Rosen. In honor of them, the issue raised by them is commonly called the EPR Effect (**vd.:** footnote). In my system set forth in my book entitled **Introspective Cosmology II**, the EPR Effect is not a particularly puzzling factor. That's because mine is a purely logical system. What does that mean?

Neither space nor extension in space are *physical*; they are *logical*. In other words, *physically* speaking, the entire universe is a singularity, which is to say a single dimensionless point. Places in space are not *physically* outside of one another in space; rather, they are only *logically* outside of one another in the order of sequence established and enforced by what I call "God's Divine Algebraic Logic Of The Mirror" (a/k/a Esoptrics, and it's a logic based on multiples and powers of 2). By the same token, the parts of a so-called "three-dimensional physical object" are not *physically* outside of one another in space; rather, they are only *logically* outside of one another in the order of sequence established and enforced by God's Divine Algebraic Logic Of The Mirror. All of that is possible because, in the final analysis, every so-called "place in space" is instead a particular kind of act which any given "generator" either currently is performing or can perform. There is no such thing as an empty place in space, save where there is a kind of act which no "generator" is currently performing. Let's now elaborate on all of that.

In my system (*i.e.:* Esoptrics), every one of the universe's "ultimate building bricks"—as some relish calling them—has two halves forever logically concentric from the standpoint of the *actuality* of each. "Carrying generator" ("generator" for short) is my system's name for one of those two halves, and "piggyback form" ("form" for short) is my system's *chief* name for the other. I emphasize "chief", because I may also refer to the forms as matrices and space envelopes.

Generators are the universe's *micro*scopic and forms its *macro*scopic ultimate constituents. That's because—if we think *figuratively*—every form can be depicted as a cube whose length, breadth, and depth may vary astronomically as it expands and contracts rhythmically; whereas, every generator can be depicted as a cube whose length, breadth, and depth are forever roughly $7.3468398 \times 10^{-47}$ centimeters (*i.e.:* 2^{-129} times the diameter of the hydrogen atom with its electron in its outermost orbit) as it "leapfrogs" its way thru the universe at one or another of an astronomical number of different frequencies.

In Esoptrics, the universe is not a collection of places in space which one or

---

[1] **NOTE OF AUGUST 14, 2009:** It now seems I've been confusing two terms. "EPR *Effect*" stands for "Enhanced Permeability and Retention Effect" and has to do with some molecules in tumors. It's "EPR *Paradox*" which refers to the Einstein-Podolski-Rosen issue. Einstein called it "spooky action at a distance" (Wikipedia: **EPR Paradox**). Physics has long wrestled with the mind-boggling suggestion that space's occupants—though very distant from one another physically—seem at times to affect one another *instantaneously*. It's a paradox only if space is *physically* extended. For, only then is there any *physical* distance. If space and its occupants are only *logically* extended, there's only *logical* distance, and the logical is overridden wherever logic demands it. In the end, "spooky action at a distance" strongly supports Esoptrics' *logically* extended world.

more kinds of less passive realities can either occupy or traverse. Instead, it's a collection of an astronomical number of different kinds of acts which any generator can perform as it carries its piggyback form around within the confines of a second form acting as the generator's space envelope.

If you want a *figurative* explanation of that, imagine a *first* tiny cube with a *second* and much larger cube which, at its center, is centered on the *first* cube, thus producing the image of a larger cube with a smaller cube at its center. With the smaller cube always at that center, this combo is moving around within the confines of a *third* cube even larger than the *second* one. This *third*, larger cube is itself entirely composed of tiny cubes, and each of its tiny component cubes is equal to the *first* cube in length, breadth, and depth. Each of this *third* cube's tiny component cubes is both an empty place in space (*figuratively* speaking) and a particular kind of act which the *first* cube can perform. To carry its piggyback form around within the confines of the *third* cube, the *first* cube—one cube at a time—becomes one with the *third* cube's component cubes. First, the *first* cube is co-terminus with one of the *third* cube's tiny component cubes for a while; then it is suddenly co-terminus for a while with an adjacent one of the *third* cube's tiny cubes; then it is suddenly co-terminus for a while with an adjacent one of the *third* cube's tiny cubes; etc.. This, though, is not a case of the *first* cube skipping what some would call "the infinite number of fractional intermediate places in space" in order to leap instantaneously from one place in space to a *physically* contiguous one equal to the prior one in size; rather, it's a case of the *first* cube ceasing to perform one kind of act and commencing to perform one of the six (***viz.:*** left, right, up, down, front, rear) which are *logically* adjacent in the order of sequence established by the logic of the mirror. In Esoptrics, there is no such thing as either *infinitely* extended or *infinitely* divisible space. Such space always was, is, and always will be self-contradictory gibberish as manifestly self-contradictory as anything can be, and, to date, Esoptrics is by far the most successful attempt to do away with such pseudo-divine space.

Since it's very likely you didn't get a sufficient hold on all of that, let's go thru it again in more detail. Just be sure to bear in mind that we are speaking *figuratively*, which is to say we're using *Geometry's* pictures to convey what are purely abstract principles of a purely *algebraic* nature. With that in mind, we recast the picture so:

First, we must imagine a series of three-dimensional matrices, each of which is a cube composed entirely of cubes, and the cubes composing any given matrix are the same "size" as all the other cubes in all the other matrices. If you wish to think in *physical* terms, then each of them is forever roughly 10^{-47} centimeters on each side, which is to say 1Φ.

Each of these same-sized cubes can be called "primary cubes". As for the matrices, we can call them "secondary cubes". Of them, there are 2^{256} (***i.e.:*** 1.1579 x 10^{77}) different kinds which, for now, we can name: #1 thru #2^{256}. Their numbers indicate how far they were from zero actuality at creation, and this I call their *native* state. Thus: At creation, #1 was only 1 step away from zero actuality, and that is forever its native state; #2 was only 2 steps away, and that is forever its native state; #3 was only three steps away, and that is forever its native state; #4 was only 4 steps

forever logically concentric to the same one generator to which it was joined when it first began to exist. For example, if a particular generator and a #1 matrix are joined, then that generator and that #1 matrix are forever logically concentric. The universe's every ultimate constituent is thus a particular microscopic and a particular macroscopic ultimate forever joined.

The matrix logically concentric to a given generator is that generator's piggyback form, and the number of that form causes its carrying generator to be "in potency" to a second matrix which I call "the form of the generator". The "form of the generator" and "the generator's piggyback form" are two different matrices.

To say a given generator is "in potency" to a given form means that given generator is limited to activating the primary cubes within that given form's actuality. To put it another way: The form of a generator (as opposed to the generator's piggyback form) is the matrix acting as the space envelope within the confines of which the generator can operate. When a generator performs a discrete act it activates a *first* particular one of the primary cubes within the actuality of the matrix to which it is in potency. When a generator switches from one discrete act to a different one, it ceases activating that *first* particular one of the primary cubes (within the actuality of the matrix to which it is in potency), and it commences to activate one of the six primary cubes logically adjacent to that first primary cube. As it switches from activating one primary cube to another, the generator carries its own piggyback form on a journey within the confines of that *other* piggyback form to which the generator is in potency.

Whenever a generator activates a primary cube, it fills it with either a one- or a two- or a three-dimensional unit of force. That's why I call it a generator. I'll not bother to explain how it does that beyond saying this: The unit of force is the result of the drag inducing differential between the actuality and the potentiality of the generator. As I said, the *actuality* of every generator is always a cube measuring roughly 10^{-47} cm. on each side; but the *potentiality* of every generator is co-terminus with the *actuality* of the form to which it is in potency (*i.e.* the actuality of the form serving as the generator's space envelope).

Activity on the part of a matrix is very different from activity on the part of a generator. For a matrix to act means it presents—to some one or more generators—a definite number of primary cubes within itself, and to change from act to act means either to increase or decrease that number. Yes, forms can change their current distance from zero actuality but never change another property I'll not here describe.

Generators never activate more than one primary cube; and so, for a generator, there is no such thing as involving two cubes on opposite sides of the center of a matrix. Oppositely, an act on the part of a matrix always extends equally to all six of its sides.

Because as many as 2^{256} matrices may be logically concentric in a super matrix, 2^{256} generators can appear to be occupying the same place in space in the eyes of those limited to thinking in the terms of what's *physical*. Even in the eyes of those able to think in the terms of what's *logical*, those 2^{256} generators can appear to be performing the same one act simultaneously. Neither observer, though, is correct. How so?! It's because each of the generators is performing an act defined in the terms of its own discrete matrix. In other words, each given generator is performing an act defined in the terms of a matrix which is serving as the form of that given generator and that given generator only.[1]

I speak of space itself as *four* dimensional. To understand what that means, first picture a primary cube within some matrix. Next, imagine a generator is activating that primary cube. But, riding on the back of that generator is a piggyback form, and that means we have a second matrix which may be presenting a second generator with a series of primary cubes to activate. In a manner of speaking, the *first* generator has a *second* generator "in its hairdo". But, riding on the back of that *second* generator is <u>its</u> piggyback form, and that means we have a *third* matrix which may be presenting a third generator with a series of primary cubes to activate. The result?! We now have 3 generators no 2 of which are moving around in the same universe. If you prefer, say it this way: They're moving around in the same universe, but not *by means of* the same space envelope. It's as if you and I have what *appears* to be the same backyard; and yet, when I move around in mine, I'm not moving around in yours—at least not *by means of* yours.

How many levels of activity may be going on relative to the very first primary cube which the very first generator above was activating? It depends upon the number of the matrix in which that very first primary cube is found. If the number of that matrix is 2^{256}, then there can be up to 2^{256} levels of activity at what geometry depicts as the same one place in three-dimensional space; if the number of that matrix is 2^{128}, then there can be up to 2^{128} levels of activity; and if the number of that matrix is 1, then there can be only 1 level of activity. Figuratively, that means this: If my back yard includes yours as a result of mine being *much* larger than, and overlapping, yours, then, as I move thru yours *by means of* mine, there shall be no interaction between us; even though, from the standpoint of geometry's extremely misleading, three-dimensional pictures, your back yard occupies the same identical area of the three-dimensional universe as does that portion of my backyard which overlaps yours.

The result of that is this: You may observe 2 ultimates racing toward a head on collision; and yet, they'll pass right thru one another, because, in a sense, they

---

[1] **NOTE OF JUNE 4, 2008:** To make the above a bit clearer, let's try an example of what the above is trying to say. To that end, go back to the image of the boxes nesting together in such a way that all have the same center point. If the nest includes 2^{256} boxes, then there are 2^{256} carrying generators all concentric with one another and concentric with the common center of all the boxes. In our imaginations, that would suggest there's only 1 generator composed of 2^{256} generators completely fused and, therefore, all performing the same one act. That, though, is merely an error brought about by the inherent danger of trying to express *algebraic* principles by means of *Geometry's* pictures. In the *algebraic* truth of it all, each of the generators, like each of the boxes, is a discrete number of steps from zero actuality. As I *prefer* to express it: Each has a unique *ontological distance* (OD for short). To put it another way: From the standpoint of *visual experience* necessarily limited to *Geometry's* <u>three</u>-dimensional pictures, each of the 2^{256} generators is in the same one place in space and, consequently, is performing the same one act; but, from the standpoint of *pure intellect's* appeal to *Algebra*, one of the generators is where it is relative to a form at OD2; another is where it is relative to a form at OD3; another is where it is relative to a form at OD4; and so forth. What to Geometry's pictures is the same one place 10^{-47} cm. LB&D is thus 2^{256} different places, and that's how the same one *seemingly* three-dimensional place in space can have a *fourth* dimension.

are not traveling in the same universe. As I prefer to put it: They are not traveling thru the universe at the same level of that huge void's fourth dimension.

D: DARK MATTER:

Another result of that is dark matter. How so?! Light emitted by bodies ostensibly in our universe may not reach us because—from the standpoint of Algebra's logic—that light is not traveling in the same universe as is our planet. More precisely, it is traveling in our universe but at a level of the void's fourth dimension which is too far above the level being used by our planet's carrying generators.

For myself, it's as simple as pie; and yet, some people are so obsessed with thinking in the terms of Geometry's three-dimensional pictures, my algebraic logic is as unintelligible to them as Relativity is to a new born.

E: WAVES VS. PARTICLES:

"Haasian" cosmology also rather easily explains how ultimates sometimes behave like particles but sometimes as waves. How so?!

If a generator is traveling thru space utilizing the primary cubes of a matrix which is itself trapped within a super matrix, the generator behaves like a particle; but, if the generator is traveling thru space utilizing the primary cubes of a matrix which has been expelled from a super matrix, the generator behaves like a wave, because it's traveling thru space utilizing the primary cubes of a matrix which, in this "free" state, requires the generator to bob outward from, and inward toward, the center of the matrix.

F: CONSCIOUSNESS:

The idea of *logically* extended space and material objects may seem ridiculous to many, because we *experience* space and material objects as *physically* extended. But, **_why_** do we experience them as physically extended? Is it because that's the way they are? By no means! Their ability to *appear* physically extended is due to the two-phased and intermittent way in which acts of consciousness occur.

That means this: As with virtually everything in the universe, every act of consciousness involves a well-defined, discrete act on the part of a matrix, and this act has a definite duration relative to all other acts on the part of all other matrices. To be more specific, every act of consciousness is an act on the part of a matrix whose native and current state is somewhere between 2^{155} (*i.e.:* 4.5671926×10^{46}) and 2^{156} (*i.e.:* 9.1343852×10^{46}). Each 2 phase cycle of that matrix endures roughly 10^{93}K (*i.e.:* 2 x the square of 4.5671926×10^{46} = 4.1718496×10^{93} and 2 x the square of 9.1343852×10^{46} = $1.66873986 \times 10^{94}$). Since Esoptrics says 1 second = 1.38847×10^{95}K, each 2 phase cycle of such a matrix would last somewhere between roughly an eighth of a second (*i.e.:* .1202 sec.) and roughly one-thirty-third of a second (*i.e.:* .03005 sec.). Such a matrix would thus perform somewhere between 8.32 and 33.28 cycles per second.

If I remember correctly, it's been well established by observation that we humans cannot detect what takes less than about a twentieth of a second (*i.e.:* .050 sec.) to transpire. For example, an old note of mine says that, on page 72 of vol. 23 of the 1967 edition of ***Encyclopedia Britannica***, we're told that, when extremely bright flashes of light are played on the eye, the eye is totally unable to detect the flashes once the rate reaches approximately 20 per second. If that's correct then the form we're describing here would correlate exactly with that. It would also correlate exactly with Esoptrics' contention that the minimum unit of time is 7.20217×10^{-96} seconds.

Such a matrix would be a cube somewhere between 2^{156} and $2^{157}\Phi$ in length, breadth, and depth, and that works out to be somewhere between 6.71 and 13.42 cm., which is to say somewhere between 2.64 and 5.28 inches. Such a form would thus encompass an area of the brain somewhere between 382.1 and 2,416.9 cu. cm., which is to say somewhere between 18.4 and 147.2 cu. in.. That fits in well with the human brain's size. Here again, then, the matrix we're describing correlates with what is observed—in this case, correlates with what is observed regarding the human brain's size.

Compare the duration of such a form's two act cycle (*i.e.:* 4.1718496×10^{93}K) with the duration of the two act cycle of a form native to, and still at, 1 step from zero actuality. That latter duration is only 4K. In other words, for every cycle on the part of matrix #2^{155}, there are 10^{93} cycles on the part of matrix #1. From the viewpoint of matrix #1, every cycle on the part of matrix #2^{155} is roughly 10^{93} times longer in duration than what matrix #1 would call roughly a 20th of its version of a second; therefore, what matrix #2^{155} experiences as taking place in the flash of a 20th of a second, matrix #1 experiences as taking an astronomical number of millennia to run its course (*i.e.:* by our reckoning, 10^{92} seconds = 10^{82} millennia).

But how can matrix #2^{155} experience a duration that astronomically lengthy and experience it as being quite brief? It's because, as long as matrix #2^{155} is performing the same one act, it undergoes no internal change whatsoever and, consequently, is aware of no kind of change whatsoever. With no awareness of any kind of change whatsoever, it has no clue whatsoever to the duration of its act; and so, its act seems, to it, wholly without duration. As you know, if you sleep soundly, 8 hours go by in a flash.

How, then, does matrix #2^{155} experience change? Like all matrices (*i.e.:* forms), matrix #2^{155} performs 2 kinds of acts: an expanded act followed by a contracted act. While performing its expanded act, matrix #2^{155} is aware of what's frozen and unchanging in the primary cubes its embracing; whereas, while performing its contracted act, it is aware of nothing. While it's performing its contracted act, an immense series of generators each act an astronomical number of times and, thereby, deposit units of force in each of the $(2^{156})^3$ (*i.e.:* $2^{468} = 7.62145 \times 10^{140}$) cubes which matrix #$2^{155}$ will experience when it performs its next expanded act.

Performing that next expanded act, matrix #2^{155} is not observing what *is happening* but, rather, the *residue* of what *happened* in a past which, from matrix #1's point of view, happened over the course of an astronomical number of millennia. That residue, of course, is spread out over primary cubes only *logically* outside of one another; but, all matrix #2^{155} can observe is that they are *outside of one another* and, not being able to observe the *manner* in which they are outside of one another, merely assumes they are *physically* outside of one another in space. Shades of Immanuel Kant!

Incidentally, according to my calculations, every primary cube is $7.3468398 \times 10^{-47}$ cm. according to our way of observing and calculating; and so, an area of one cubic cm. would hold $(1.3611295 \times 10^{46})^3 = 2.5217286 \times 10^{138}$ primary cubes. Matrix #2^{155} would contain roughly 3,000 times that number of primary cubes (*i.e.:* $2^{468} = 7.62145 \times 10^{140}$). That would allow its acts of consciousness to view a very large area with a very high level of resolution. It would, of course, be seeing a highly miniaturized version of the original.

Finally, how does matrix #2^{155} live in and experience a world radically different from the world matrix #1 lives in and experiences? In other words, how can the latter be, not merely a part of the former but, rather, a subservient part of it? The answer is this: Matrix #2^{155} lives in and experiences a world populated by other matrices each having a number somewhat close to its own. Matrix #2^{155} is thus a kind of strobe which causes the rapidly rotating fan blades to become stationary blades and, thereby, matrix #2^{155} experiences what's going on from the standpoint of matrices changing at or near its own rate of change.

As for how matrix #1 is a subservient part of the world of matrix #2^{155}, bear in mind that matrix #1 encompasses only 8 primary cubes; whereas, matrix #2^{155} encompasses roughly 10^{140} primary cubes. It's merely a case of the *macro*scopic exercising its influence over far more of the *micro*scopic than the microscopic itself can.

Science, these days, puzzles over how the cloudy and fitful world of microscopic particles gives birth to our clear and reasonably orderly macroscopic world.[1] The problem here, of course, is that Science simply knows virtually nothing about matrices (*i.e.:* forms) and, therefore, virtually nothing about the universe's *macro*scopic ultimate particles and how they connect to, and govern, the universe's *micro*scopic ultimate particles. Only Haasian cosmology really throws any light on that issue, and the light it throws is far from paltry.

Yes, only Haasian cosmology details the *macro*scopic ultimates and how they connect to, and govern, the microscopic ultimates. For all of that, to this cosmology, both scientists and philosophers have—for 45 years now—turned an ear as deaf as any ear can be. Let them *continue* to do so, and let us see what history shall one day say of them.[2]

---

[1] **NOTE OF AUGUST 12, 2009:** See, for example pg. 43 of John Polkinghorne's **Quantum Theory, A Very Short Introduction**, Oxford Press, NY, 2002. There, he basically admits this: If we rely solely upon the universe's "quantum constituents (*i.e.: micro*-scopic entities such as quarks, gluons, and electrons) to explain the world of large objects we daily observe, then we have a tremendous problem, namely: The behavior of the "quantum constituents" is too fitful and cloudy to give rise to our daily world's *macro*-scopic entities.

[2] **NOTE OF AUGUST 12, 2009:** The behavior of the *micro*-scopic quantum constituents is too fitful and cloudy to give rise to our daily world's "biggies". The universe simply *cannot*—as Science has long *assumed*—be *nothing* but *micro*-scopic *quantum* entities; instead, It *must* be that, among its *ultimate* constituents, the universe includes a host of *macro*-scopic *ultimates* which are really present no matter how unobservable they might be. I, therefore, insist that the future of humanity's control

G: EXCHANGES WITH MY BROTHER GORDON:

G-1: EXPLICIT VS. IMPLICIT:

I should quit now. For all of that, permit me to throw in here what, this very day, I sent to the youngest of my three brothers. When I sent him an e-mail about how my system answers the problem of the EPR Effect, he fired off a "quickie" e-mail containing a "quickie" criticism of my appeal to a logically extended universe. I fired back so:

What?! You said: 'the logical is filled with "implicit" consequences and not explicit ones'?! That's a strange statement to say the least. "Implicit" versus "explicit" has to do with the difference between what is merely *implied* rather than *expressly* stated. The logical versus the physical has to do with what is necessary according to the principles of right reason versus what is necessary according to either observation or the principles of mechanics. There's no correlation whatsoever between implicit/explicit and logical/physical.

When we say that 2+2 = 4, we utter a very *explicit* statement which is necessarily true according to the principles of sound reasoning. That being the case, it is irrelevant to talk about whether or not it *implies* more than it's *expressly* stating. Conversely, when we say decapitation causes death, we utter a very *explicit* statement which is necessarily true according to the principles of mechanics. And since when does such a statement not *imply* anything more than what's *expressly* stated? And since when does its importance to us depend upon whether or not it *implies* more than it *expressly* states? By the same token, when, on a cloudless day at noon, we say that the Sun is in the sky, we utter a very *explicit* statement which is necessarily true by observation. Here again, is there absolutely nothing *implied* over and above what's *expressly* stated, and should the statement be dismissed if it can be taken to *imply* more than it *expressly* states? Dear God! Have pity!

Of course, you could be saying that, the instant one speaks of what is merely *logically* so, scientists immediately fear this is Philosophy, and Philosophy always *implies* far more than anyone can possibly foresee; and so, it scares them; whereas, in the eyes of the scientists, what is *physically* so is transparent, which is to say it *implies* nothing more than what they can observe staring them in the face. In that case, I would say this: Such scientists are suffering from that mental illness

---

of its environment lies chiefly in humanity's ability to forsake Science's fixation upon the universe's <u>micro</u>-scopic ultimate constituents, to go, instead, in search of its <u>macro</u>-scopic ultimates, and, finally, to gain exhaustive knowledge of those <u>macro</u>-scopic ultimate constituents. Who shall most effectively pursue that goal: the scientists or the philosophers? It's easy enough to see how far this philosopher has gone in that direction. To be sure, I've only scratched the surface and have done so little I can make no significant predictions. Still, I dare boast that no scientist has come close to saying as much as this philosopher has regarding what the universe's <u>macro</u>-scopic ultimates might possibly be like and how they might possibly affect our world. When our need for knowledge regarding those macroscopic ultimates is as great and obvious as it is—and Science's inability to answer that need as great and obvious as it is—it is, I suggest, a crime against humanity to ignore what this philosopher has to say about the universe's <u>macro</u>-scopic ultimates.

commonly called paranoia.

Do you perhaps mean to tell me that scientists never go by what *logic* says is necessary? How, then, is there such a thing as *theoretical* physics? I'll tell you how: Scientists say: "We have observed such-and-such, and, according to the principles of sound reasoning, it must be that such-and-such a kind of particle must exist. We can't *observe* that it exists; but, it *must* exist because logic demands that, given what we do observe, it *must* be so."

G-2: LOGICALLY VS. PHYSICALLY DISTINCT:

Finally, maybe you're trying to say this: Whatever is only *logically* distinct and separate from another is not *really* distinct and separate from it. You are, of course, quite within your rights to maintain such; but, just be sure to bear in mind that it is merely a *contention* on your part, and that, of course, is my whole point, namely: All Science—if it obstinately clings to the *contention* that all separation and distinctness must be *physical* in order to be real—is clearly *quack* Science, and one of the reasons why it is clearly *quack* Science is that the EPR Effect gives the lie to "all real separation is *physical* separation" and trumpets the truth that "real separation is indeed *logical* separation."

I say it another way: To be *really* distinct and separate from one another, things need only have a different set of internal characteristics and do not by any means need to be *physically* separate from one another in space and/or time. *Real* separation and distinction stem from a thing's ability to show an intelligent observer that it has its own unique set of internal characteristics, and, therefore, does not by any means stem from the ability of some infinitely extended, three-dimensional blob of nothingness in the middle of nowhere to give each separate thing its own unique place to occupy within the confines of that real hole in real nothingness. (Imagine that! Purely logical separation is not real; but, separation by means of nothingness is. What a croc of the well-known stuff!) Apparently, though, no matter how loudly the EPR Effect screams the truth, there are some who will never hear the truth come Hell or high water.

G-3: SCIENCE'S LEVEL OF EXPLICITNESS:

My brother replied to the above with a more detailed e-mail. I countered with this:

Let's try this yet another time. In keeping with what seems to be your preferred terminology let's invoke the word "explicitness" and say it shall hereafter mean this: "the capacity of a system of thought to produce mathematically precise equations which, in a laboratory setting, can be shown to predict and describe accurately how some particular target of observation shall affect our measuring instruments." With that definition mastered, I would then dare to suggest you're trying to tell me this: "Every system of repute in the scientific world is one which has explicitness; whereas, yours does not, at least not yet."

To that, I reply: So what?! After all, it's almost universally admitted that every one of the systems touted by the scientific world has only a *partial* explicitness, which is to say that, in the scientific world, there is not a single, solitary system which has explicitness save with regard to a small part of the whole. To support that last contention, I now give you the words of Stephen Hawking on pages 11 and 12 of his book ***A Brief History Of Time*** as published in paperback form by Bantam Books, New York, 1998.

> It turns out to be very difficult to devise a theory to describe the universe all in one go. Instead, we break the problem up into bits and invent a number of partial theories. Each of these partial theories describes and predicts a certain limited class of observations, neglecting the effects of other quantities, or representing them by simple sets of numbers. It may be that this approach is completely wrong. If everything in the universe depends on everything else in a fundamental way, it might be impossible to get close to a full solution by investigating parts of the problem in isolation. Nevertheless, it is certainly the way that we have made progress in the past.

By way of example, Newton's system allows us to predict exactly where which planet shall be when. Who would say, though, that his system gives us more than a pittance relative to all we need to know about the universe? Indeed, who would deny that his system falls on its face the instant you go beyond a rather limited area of questions regarding the universe? No system of thought in the scientific world, however much explicitness it has, fails to fall on its face at some point. Here again, I give you Hawking's words from pg. 12 of the above-mentioned work:

> Today scientists describe the universe in terms of two basic partial theories—the general theory of relativity and quantum mechanics. They are the great intellectual achievements of the first half of this century . . . . [Skip] Unfortunately, however, these two theories are known to be inconsistent with each other—they cannot both be correct.

That's right. Every one of the scientific community's "great intellectual achievements" is a "partial theory" which fails "to get close to a full solution". That being the case, we can point to any and every one of those "great intellectual achievements" and say: "Its explicitness is so paltry relative to the whole, its explicitness *proves* nothing save ***partial utility***. By no means does its explicitness *prove* either that it is correct or that it applies to the *real* world." Here again, I give you the words of Stephen Hawking. On this go around, we turn to his book entitled ***The Universe In A Nutshell*** in its hardbound version as published by Bantam Books, New York, 2001. First, on page 59, he writes:

One might think this means that imaginary numbers are just a mathematical game having nothing to do with the real world. From the viewpoint of positivist philosophy, however, one cannot determine what is real.

Next, on page 31, he writes:

> Any sound scientific theory, whether of time or of any other concept, should in my opinion be based on the most workable philosophy of Science: the positivist approach put forward by Karl Popper and others. According to this way of thinking, a scientific theory is a mathematical model that describes and codifies the observations we make. A good theory will describe a large range of phenomena on the basis of a few simple postulates and will make definite predictions that can be tested. If the predictions agree with the observations, the theory survives that test, *though it can never be proved to be correct* [Emphasis mine – ENH]. . . . [Skip] If one takes the positivist position, as I do, one cannot say what time actually is. All one can do is describe what has been found to be a very good mathematical model for time and say what predictions it makes.

Yes, he can't say "what time actually is"; but, the Haasian system can: It's the measure of the rate at which an absolutely ultimate constituent of the universe changes from one specific primary act to a different one. Since ultimates can do that at 2^{385} (*i.e.:* 7.88×10^{115}) different rates, time is relative to the ultimate at hand and "flows" ("pulses along" would be more accurate) at 2^{385} different rates ($2^{385} = 2 \times [2^{128}]^3$). But, back to the mainstream!

What Hawking says above is echoed by John Polkinghorne in the little booklet you sent me (How can I thank you enough for sending it?). On pages 82 and 83, he writes:

> Positivists see the role of science as being the reconciliation of observational data. If one can make predictions that accurately and harmoniously account for the behaviour of the measuring apparatus, the task is done. Ontological questions (What is really there?) are an irrelevant luxury and best discarded. The world of the positivist is populated by counter readings and marks on photographic plates. . . . [Skip to pg. 83.] Niels Bohr often seemed to speak of quantum theory in a positivistic kind of way. He once wrote to a friend that

>> There is no quantum world. There is only abstract quantum physical description. It is wrong to think that the task of physics is to find out how nature *is*. Physics is concerned with what we can say about nature.

On pages 46 and 47, Polkinghorne writes:

One argument in favour of this stance is the positivist assertion that science is simply about correlating phenomena and that it should not aspire to understanding them. If we know how to do the quantum sums, and if the answers correlate highly satisfactorily with empirical experience, as they do, then that is all that we should wish for. It is simply inappropriately intellectually greedy to ask for more.

So then, the explicitness of any and every one of Science's "great intellectual achievements" is so partial and paltry, it leaves even scientists of Hawking's stature wondering if these "great intellectual achievements" have anything to do with reality or can be proven correct. With explicitness that universally funky, what's the point in demanding it of any theory before you decide whether or not it deserves examination?

G-4: AN ALTERNATIVE TO EXPLICITNESS:

In point of fact, Hawking himself admits that's not *always* the way to go about determining whether or not a theory should be examined. Turn to page 141 of *A Brief History Of Time*. There, he writes:

> I'd like to emphasize that this idea that time and space should be finite "without boundary" is just a *proposal*: it cannot be deduced from some other principle. Like any other scientific theory, it may initially be put forward for aesthetic or metaphysical reasons, but the real test is whether it makes predictions that agree with observation. This, however, is difficult to determine in the case of quantum gravity, for two reasons . . . . [Skip] Second, any model that described the whole universe in detail would be much too complicated mathematically for us to be able to calculate exact predictions. One therefore has to make simplifying assumptions and approximations—and even then, the problem of extracting predictions remains a formidable one.

Yes, indeed! Haasian cosmology, far more so than any other system, describes "the whole universe in detail" and is, therefore "too complicated mathematically for us to be able to calculate exact predictions." With someone of Hawking's stature admitting that, why is my system totally ignored, and totally ignored now for 45 years? It's because of this damnably irrational and duplicitous *fixation* upon explicitness despite the fact—as positivist scientists themselves admit—that there is nowhere in Science's "great intellectual achievements" an explicitness which is anything more than a joke when it comes to *proving* what *is*.

Absolutely, there *has* to be a *second* criterion by which to judge the merit of a system, and it's quite obvious to me what that other criterion must be—namely: whether or not it gives a viable explanation of how the universe could possibly produce a phenomenon which, until now, evades every attempt to explain how it could possibly occur. With its explanation of how the EPR Effect could possibly occur, my system meets that criterion to a degree which should boggle the mind of every individual *rightly* called a scientist.[1]

G-5: SCIENCE & THE WHAT VS. PHILOSOPHY'S FAILURE TO PROVIDE THE WHY:

Finally, let's turn again to *A Brief History Of Time*, and consider these words from Hawking on pages 190 & 191:

> Up to now, most scientists have been too occupied with the development of new theories that describe *what* the universe is to ask the question *why*. On the other hand, the people whose business it is to ask *why*, the philosophers, have not been able to keep up with the advance of scientific theories. In the eighteenth century, philosophers considered the whole of human knowledge, including science, to be their field and discussed questions such as: did the universe have a beginning? However, in the nineteenth and twentieth centuries, science became too technical and mathematical for the philosophers, or anyone else except a few specialists. Philosophers reduced the scope of their inquiries so much that Wittgenstein, the most famous philosopher of this century, said, "The sole remaining task for philosophy is the analysis of language." What a comedown from the great tradition of philosophy from Aristotle to Kant!

Obviously, Professor Hawking has never heard of Edward N. Haas. I wonder why. Yes, it is for *philosophers* to tell us *why* the universe is the way it is, and Edward N. Haas, a philosopher, has done that to an extent which should leave all the world breathless. For all of that, his efforts have met only with 45 years of *total* indifference.[2] After all, it doesn't have that one awesome characteristic which other systems possess to an amazing degree, namely: *explicitness* (*i.e.:* the ability to make predictions). As I said in an earlier E-mail: What a croc of the well-known stuff! One of these days, O scientific community, you shall answer to history for this total indifference, and it won't be a pretty day for you when it does happen.

Oh, how you aggravate me, you who are so obstinate in your refusal to pay

---

[1] **NOTE OF AUGUST 18, 2009:** If, as Esoptrics contends, there is only *logical* extension and no such thing as *physical* extension, then "spooky action at a distance" ceases to be *spooky* to any extent whatsoever. Because its concept of *logical* distance so thoroughly explains how action at a distance is possible, Esoptrics gives powerful evidence that it deserves examination.

[2] **NOTE OF AUGUST 15, 2009:** Perhaps one can explain this total indifference by giving Hawking back a paraphrase of his own words like so: "Scientists have not been able to keep up with the advances in Philosophy. In Edward N. Haas, Philosophy has simply become too technical and mathematical for the scientists or anyone else except for Edward N. Haas himself."

heed to the ideas given you thru me! For all of that, I still close wishing:

May the Lord bless you and keep you,
And lift up His Countenance upon you.
Yea, may He make His Face to shine upon you,
And give you Peace.

In Christ Jesus,
        Thru Mary
                With Joseph,
A Brother To All:

EDWARD N. HAAS

שָׁלוֹם

POSTSCRIPT OF OCTOBER 5, 2009: In an e-mail dated 8/21/07, my brother Gordon wrote:

> I totally agree it [time] is a measure of change, but it surely doesn't need change to exist, change is an inherent byproduct. That's all I am saying. And it seems somewhat invalid to approach it from the reverse side.

My brother has it backwards. Time, as we know it, is the "inherent byproduct" of change, and not vice versa. *The duration of the totally changeless* doesn't need change to exist; but, certain it is that, if there is some kind of time which measures the duration of the changeless, we—we who experience and know only the kind of time which measures change—cannot possibly know the slightest thing about it other than that it somehow measures the duration of the changeless. From of old, perhaps the vast majority of philosophers and theologians have admitted that one of the primary differences between our world and God's is that the latter is above time and is above time precisely because it is above every shadow of change (***vd.:*** St. Thomas Aquinas' ***Summa Theologica, Pars Prima***, Question 10, Article 4). As a result, for them, one of the most puzzling of issues has ever been: "How much time passed before God created our world of change?" Their only answer was ever: "None!" Esoptrics answer is: "None of any kind to any extent known to us!" It is, then, my brother who is invalidly approaching this issue from what has, for millenia, been "the reverse side" in the eyes of an ocean of philoso-phers and theologians.

The historian finds that great events, even the most important changes in the commercial relations of the world, . . . had their origin not in the combinations of statesmen, or in the practical insights of men of business, but in the closets of uninterested theorists, in the visions of recluse geniuses. . . . All the epoch-forming revolutions of the Christian world, the revolutions of religion, and with them the civil, social, and domestic habits of the nations concerned, have coincided with the rise and fall of metaphysical systems.

——**SAMUEL TAYLOR COLERIDGE.** As quoted by R. J. White on pages 15 & 16 of ***Political Tracts of Wordsworth, Coleridge, and Shelley***, London, 1953.

$\mathcal{D}$ocument #4:

LETTER STARTED JUNE 8, 2007:

A. OUTSIDE OF TIME VS. OUTSIDE OF *CONTINUOUS* TIME

Prof. D . . . . ., Chair
Dept. Of Philosophy
University of . . . ., NY

Greetings, Professor D, my dear, kind and generous Sir:

A thought crossed my mind, saying: "It's liable to take quite some time to answer *each and every* point he raised in his e-mail of Monday, June 4, 2007, 17:52:05. Therefore, if you send him nothing until you've produced a letter addressing *each and every* point, it may be weeks before you send him anything. Why not send him multiple letters each limited to addressing only one of the points he raised? That way, you'd be able to send him something rather quickly." It's a thought I found quite acceptable; and so, here is a first letter—a letter addressing what you repeatedly refer to as the "a-temporal nature"[1] of my "basic entities".

Almost at the beginning of your e-mail, you wrote:

> Your basic hypothesis seems to be that basic realities do not exist in space and time but are outside of space and time. This is the view of Leibniz and other members of the 17th c. rationalists school: Leibniz held that the basic entities, or monads, are unextended centers of energy, and that space and time are phenomenal, pertaining to appearance, not real.

That is far from being my hypothesis. For me, the issue was never whether or not basic realities are outside of *time*; it was always whether or not they are outside of *continuous* time. I say yes, because I say *continuous* time is "phenomenal, pertaining to appearance, not real". I'm far from saying that of *time itself*. How could you or anyone else fail to notice that in a flash? What, then, must I say to overcome this strange inability to grasp my "basic hypothesis"—basic hypothesis, that is, regarding the relationship between "basic realities" and time?

---

[1] Actually, in most instances, you wrote "atemporal", and, only near the end, did you throw in the dash and write "a-temporal".

B. ORIGIN OF THE TERM "ESOPTRICS":

Before tackling that task, let's briefly touch on a somewhat childish issue. The instant I use phrases such as "my basic hypothesis", "my system", "my theory", "Haasian Cosmology", and the like, some immediately thunder: "Egomaniac!" Oh yes! There's no question the charge of "Egomaniac" is 1,000% true in my case; but, one must give others as few excuses as possible to meander off onto tangents which only waste valuable time.

To that end, let's replace the above egomaniacal terms with "Esoptrics". It comes from the Greek word *"esoptron"*(εσοπτρον)[1]. As you may well know, that's the Greek equivalent of the English word "mirror". Since the basis of all the following has to do with the way mirrors work (*i.e.:* doubling and reversing), "Esoptrics" is a very appropriate choice. Let us talk, then, about Esoptrics, the Divine Algebraic Logic Of The Mirror.

C. PIGGYBACK FORMS & THEIR CARRYING GENERATORS:

By now, you are probably aware that Esoptrics distinguishes between two major classes of what you call "basic realities"—entities Esoptrics generally refers to as "ultimate constituents". On the one hand, Esoptrics speaks of the universe's _micro_scopic and, on the other, of its _macro_scopic ultimate constituents. The former Esoptrics calls "generators" and "carrying generators"; the latter, it calls "piggyback forms" and "matrices". Every piggyback form's every act is concentric with the act of a carrying generator eternally joined to it. That's why the former is called a *piggyback* form and the latter a *carrying* generator.

Nowhere does Esoptrics suggest either of these ultimates is outside of either time or space. Because of what I said in my opening paragraph, I shall, for now, ignore the issue of space and dwell on that of time.

D. OUTSIDE OF TIME = INTERMITTENTLY OUTSIDE OF CHANGE & OUTSIDE OF THE KIND OF TIME WHICH MEASURES CHANGE:

Esoptrics suggests that the universe's ultimates are *intermittently* outside of time. More specifically, it suggests they are intermittently outside of time *as we experience time*. Immediately, that raises the question: Just how _do_ we experience time?

Patently, I cannot speak for everyone who has ever lived. I can only speak for myself, for those whose writings on time I've read, for those of my many relatives, friends, and acquaintances with whom I've discussed time, and for those "talking

---

[1] In the **New Testament**, *"esoptron"* is the spelling used by St. James (**James** 1:23) and St. Paul (**I. Cor.** 13:12). The **Greek-English Lexicon** of Lidell and Scott adds in an additional "i" and spells it "eisoptron". Pardon me if I cling to Sts. Paul and James.

heads" in the media who have publicly commented on the issue. What, then, do all of these sources tell me about time?

Time, they all say without exception, has to do with *change*. As some express it: Time is the measure of change. It is *change* which is the *sine qua non* (*i.e.:* the absolutely indispensable ingredient) of time. Therefore, for me and the many I know, it is unanimous that there is no such thing as time, save where there is:

CHANGE

FROM

what, before the change, was current

TO

what, *before* the change, was in the future and, *after* the change
is—in some way and to some extent—different from what preceded the change.

E. THE PRIMARY VS. THE SECONDARY ACTS OF FORMS & GENERATORS:

What, though, does Esoptrics say of change? First, it says this: Whether microscopic or macroscopic, each of the universe's ultimate constituents is—in each instance of its most basic act (*i.e.:* what some philosophers would call its "first act" or "primary act" and "act of existence")—a very unusual kind of act. Here, you see, we are talking about that *primary* act of *existence* which serves as the means by which this or that particular form or generator manages to be the particular ultimate which it is.[1]

For Esoptrics, this *primary* act of *existence* is a very unusual kind of act. It's so unusual, this *primary* act is to no extent similar to any of the actions we observe either on the part of the objects we experience with our senses or on the part of the molecules, atoms, and sub-atomic particles we detect with our instruments.

All such latter actions are merely *secondary* acts on the part of the ultimate constituents, or on the part of the particles, atoms, molecules, etc., which the ultimates produce by combining. Such *secondary* acts are the *behavior* (as opposed to the *existence*) of whatever performs them. In these *secondary* acts, we witness the effects produced by a series of successive *primary* acts (*i.e.:* acts of existence) on the part of one or more generators manipulated by their piggyback forms—manipulated much as puppets are manipulated by their puppeteers. As St. Thomas Aquinas' supreme insight might say it: Only in The Infinite are existence and essence (*i.e.:* primary and secondary acts) the same; elsewhere, they differ radically.

---

[1] **NOTE OF DECEMBER 21, 2008:** Existence is the "means by which" a given reality is whatever it is, and essence is "what" a given reality is, which is to say essence is the sum total of a given reality's indispensable internal characteristics plus its indispensable behavior patterns.

With regard to this primary act of existence, one of Esoptrics' most crucial hypotheses is this:

AS LONG AS A GIVEN ULTIMATE IS PERFORMING THE SAME, ONE, SPECIFIC PRIMARY ACT OF EXISTENCE, IT'S PERFORMING AN ACT WHICH UNDERGOES NO KIND OF INTERNAL CHANGE WHATSOEVER.

Most assuredly, it is difficult for us to imagine how there could be such a thing as an **act** which involves no kind of internal change whatsoever. That, though, is because it's almost impossible for us to appreciate just how vast is the difference between primary and secondary acts. On that point, we are like someone blind from conception trying to imagine what it's like to see a visual image. Because only *infinity's* occupants can experience the internal characteristics of an ultimate's *primary* act of *existence*, no occupant of this *finite* world can even begin to imagine either what it's like to experience such an act or how such an act could be devoid of internal change throughout its course.

Recall, too, what an immense cloud of theologians—at least two thousand years deep—has daily thundered in our ears: "God is a single, unchanging act." If there can be no such thing as an *act* devoid of change, then neither can there be One rightly called God.[1]

Nevertheless, regardless of what we can or cannot "picture" in our heads, every individual primary act on the part of a generator or a piggyback form involves no kind of internal change whatsoever. Ultimate constituents are **NEVER INVOLVED IN INTERNAL CHANGE**, save where they change from one specific *primary* act to a different one. But, the *sine qua non* of time is **CHANGE**; therefore:

NO ULTIMATE CONSTITUENT OF THE UNIVERSE IS EVER INVOLVED IN TIME, SAVE WHERE IT CHANGES FROM PRIMARY ACT TO PRIMARY ACT.

Most certainly, clearly, and necessarily: As long as *change* is time's absolutely indispensable ingredient, there can be no such thing as time for any and every specific ultimate *while it's performing an act devoid of change*. More exactly, there can be no such thing as *involvement* in time for any such specific ultimate. Clearly, though, that's an *intermittent* lack of involvement in time.

F. HISTORY'S TWO STREAMS OF

---

[1] According to Catholic teaching, every occupant of infinity continuously performs a single act in which there is never the slightest trace of any kind of change. If so, how do infinity's occupants experience the duration of their one act? I dare speculate they are aware of a kind of time the likes of which no *finite* being can even begin to imagine. If there is such a kind of time, it's a kind of which we know nothing save that it is nothing like the kind of time we experience and that it measures the duration of a changeless act. For us, "time" means only *experienced* time, and only *experienced* time has any relevance to our world, and such time is found only where there is change.

TWO RADICALLY DIFFERENT KINDS OF MOMENTS:

Is that not a proposal as clear and simple as proposals ever get to be? For all of that—although I have been shouting this from the rooftops for *forty-five years*—I don't think there has ever been a single, solitary hearer who has understood what I am saying. Why is that?

Esoptrics *suggests* this answer: *Apparently*, all but myself, *assume*—assume most blindly and tenaciously—that history involves only a **single** stream of only **ONE** kind of moment: moments of time. By "history", of course, I mean the universally *inexorable* journey from past to future (I didn't say universally *continuous* since Esoptrics says no kind of change ever occurs below roughly 10^{-95} of our seconds.), and Esoptrics views this "history" in a manner apparently limited to Esoptrics and Esoptrics alone. Esoptrics—as you might expect from the logic of a *mirror*—says that history involves $\mathbf{2^{385}}$ (**i.e.:** 7.88×10^{115}) streams each of which is **TWO** radically different kinds of moments alternating with one another. They are:

DURATIONLESS MOMENTS OF TIME
VS.
TIMELESS MOMENTS OF DURATION.[1]

The relationship between these two kinds of moments can be *geometrically allegorized* (**i.e.:** *figuratively* represented by a drawing) by a series of dashes alternating with a series of dots like so:

——— . ——— . ——— . ——— . ——— . ——— . ——— . ——— .

We can refer to the above drawing as a "history stream", or, if you prefer, a "cosmic history stream". That's because the drawing seeks to express how the universe moves from its past to its future.

In the above history stream, each dash represents a moment of timeless duration, and each dot represents a moment of durationless time. More precisely, each dash represents a particular primary act on the part of one of the universe's ultimate constituents. Esoptrics, of course, says these ultimates are either generators or piggyback forms;[2] and so, Esoptrics says each dash represents a particular primary act on the part of either a generator or a piggyback form. As long as a given ultimate is performing the same, one primary act, it's doing what involves no kind of internal change whatsoever, and, since time is the result of, and measure of, change, the ultimate has no involvement in time as long as it does not change from one particular primary act to another one.

---

[1] In Chapter Thirty-Eight of **Introspective Cosmology II**, I explain how this introduction of two radically different kinds of moments annihilates Aristotle's argument against time composed of indivisible instants. I'll not go into the details here. They're given in document #12.

[2] According to Esoptrics, a particle such as a quark is a massive number of generators and piggyback forms made concentric to one another when a piggyback form accelerates.

Obviously, that leaves the door open to the possibility of an ultimate changing from one particular primary act to another one, and, as often and long as it does so, it is involved in time. But, changing from one primary act to another is something which occurs *instantaneously*, because, between the end of one primary act and the start of the next, there is no interval or pause of any kind. As a result, moments of time (*i.e.:* moments of involvement in change) are durationless. That, then, is why Esoptrics refers to moments of time as "durationless".

G. 2^{385} HISTORY STREAMS

Esoptrics says there are 2^{385} such "history streams". How is that possible?

First of all, let's be more specific about what the term "history" (or "cosmic history", if you prefer) means to Esoptrics. Esoptrics says that, just as there are nine choirs of angels, the universe is currently in its ninth epoch; and so, for Esoptrics, "history" currently means the progress of the universe from the beginning to the end of its ninth epoch—a progress wholly and universally on hold below roughly 10^{-95} of what we experience as a second.

But, the word "universe" is our way of referring to an astronomical number of ultimate constituents taken as a whole. That being the case, "history" can also refer to any and every one of those ultimates. Where it does so, it denotes the progress of one or more ultimates from the beginning to the end of the ninth epoch.

Esoptrics says that—throughout this the ninth epoch of the universe—the frequency with which ultimates change from primary act to primary act can vary. Esoptrics says this change from primary act to primary act can occur at 2^{385} different frequencies. Naturally, the more frequently an ultimate changes from primary act to primary act, then the more frequently do durationless moments of time occur. To put it another way:

**FOR ANY GIVEN ULTIMATE,
THE NUMBER OF DURATIONLESS MOMENTS OF TIME
OCCURRING IN ITS HISTORY
INCREASES AS DOES THE FREQUENCY OF
CHANGE FROM PRIMARY ACT TO PRIMARY ACT
ON THE PART OF THE GIVEN ULTIMATE.**

To illustrate that, let's first imagine that the left side of this page represents the beginning of the universe's ninth epoch and the right side the end of that ninth epoch. Figuratively, then, the history of the ninth epoch runs from the left side of the page to the right side. As it does so, the participants in that history journey along one or more history streams which we can figuratively represent so:

84

As you no doubt can readily appreciate, I cannot draw 2^{385} streams. Neither can we *really* do even so little as *imagine* 2^{385} streams. Nevertheless, we can imagine a rather lengthy extension of the principle we see illustrated in the above—namely: Each successive line has one dash and one dot more than the line above it. As a result, the first line has a single dash and a single dot, and the last line shall have 2^{385} dashes and an equal number of dots. Yes, there are an astronomical number of ultimates which perform only one primary act in the whole history of the ninth epoch—a history which, Esoptrics says, lasts roughly 18 trillion Earth years.

Clearly, in such a scenario, what sets each history stream apart from all the others is the discrete number of alternating moments comprising each history stream. Naturally, by "alternating moments", I mean a moment of timeless duration followed by a moment of durationless time.

H. ONTOLOGICAL DISTANCE ALSO CALLED "OD":

What determines how often a given ultimate shall change from primary act to primary act? Esoptrics answers with the concept of ontological distance—a concept it often signifies by the abbreviation "OD". What does that concept tell us?

As you no doubt know, "ontology" comes from two Greek words signifying knowledge of Being. Unfortunately, "Being" is a meaningless word to most of today's people. Perhaps, then, we'd not be too far off base to think of Being as some unusual kind of energy or power. In that case, "ontological distance" (*i.e.:* "OD") refers to a kind of distance which has something to do with Being, which is to say with some unusual kind of energy or power. How so?!

Esoptrics says God is The Infinite Plenitude Of Being and, as such, continuously makes available—to every piggyback form in the ninth epoch—a particular quantity of *potential* Being. If it helps, think of *potential* Being as a kind of *latent* power waiting to be converted into *active* power. Latent power (*i.e.:* potential Being) and active power (*i.e.:* Being = actual Being) are each truly a kind of reality; and yet, the two kinds are by no means *equally* real.

As for the quantity of this potential Being always available to every form, let's refer to it as Z. If you *must* have a *picture*, imagine each form is a mysteriously powerful center surrounded by its own mantle of perfectly transparent and powerless "ether".

If a piggyback form's primary act converts only 2^{-256} Z (Notice the minus sign.) into actual Being (*i.e.:* if it converts 2^{-256} of the universally available *latent* power into *active* power), then its ontological distance is 1, because it is only 1 step away from zero actual Being. If a piggyback form's primary act converts $2(2^{-256})$ Z into actual Being, then its ontological distance is 2, because it is 2 steps away from zero actual Being. If a piggyback form's primary act converts $3(2^{-256})$ Z into actual Being, then it's at OD3, which is to say 3 steps away from zero actual Being. Every piggyback form at OD4 is 4 steps away from zero actual Being and, therefore, its primary act converts $4(2^{-256})$ Z into actual Being. Surely you can follow the progression and see that, at OD2^{256}, the primary act of the form at that OD converts 100% of Z into actual Being (*i.e.:* its center has "swallowed" all its "ether").

According to Esoptrics, there is only one form at $OD2^{256}$ throughout the course of this the universe's 9th Epoch and, throughout that course, it performs only that same one primary act in which 100% of Z is converted into actual Being. For Esoptrics, that means this: Throughout the 9th epoch, the universe shall only expand and never contract.[1]

I. THE CORRELATION BETWEEN OD & SIZE:

Because of what's just been said, there is a necessary correlation between OD and size. How so?!

Take the *actual* Being of the form at $OD2^{256}$, and compare it with the *actual* Being of a form at OD1. The former is 2^{256} (*i.e.:* 1.15792×10^{77}) greater than the latter. Therefore, if the former be taken to encompass the entire universe and its diameter be expressed as roughly 10^{31} cm. (*i.e.:* roughly 18 trillion light years), then the diameter of the latter must necessarily be expressed as roughly 10^{-47} cm.. At OD2, we double that tiny measurement. At OD3, we triple it. At OD4, we quadruple it, and so forth.

The square root of 2^{256} is 2^{128} (*i.e.:* 3.40282×10^{38}). Therefore, even at $OD2^{128}$, we're still dealing with what's only around 10^{-8} centimeters. That's roughly the diameter of the hydrogen atom, and, for Esoptrics, the smallest quantity of actual Being in the universe's forms is 2^{-128} (Note the minus sign.) and the largest 2^{128} times the diameter of the hydrogen atom. The hydrogen atom is thus at the square root of it all.[2]

J. HOW FORMS CHANGE FROM PRIMARY ACT TO PRIMARY ACT:

Hopefully, you now have a sufficient grasp of what Esoptrics means by ontological distance. If so, we are now *almost* ready to state the rules governing how often a piggyback form shall change from primary act to primary act.

We're not *totally* ready, though, because we must first elaborate upon what's

---

[1] **NOTE OF DECEMBER 20, 2008:** The ninth epoch of the universe shall come to its close as it is roughly 18 trillion Earth years old. As it ends, the tenth epoch of the universe shall commence, creation shall be renewed, the number of ultimate constituents in the universe shall increase astronomically, and this new universe shall commence to expand to a diameter 2^{256} times the diameter of the 9th epoch's universe. In the course of that 10th epoch, the universe created in the 9th epoch shall commence to contract. However, at that point in cosmic history, the diameter of the 9th epoch's universe shall be astronomically smaller than the diameter to which the 10th epoch's universe shall be expanding. As a result, to occupants of the 10th epoch's universe, the collapsing 9th epoch's universe shall seem an insignificant event limited to what they would measure as no larger than a single atom, since, for them, every atom would include $OD1 \sim 2^{256}$.

[2] **NOTE OF DECEMBER 20, 2008:** Among the forms, the diameter of the smallest quantity of actual Being is 2^{-128} times the diameter of the actual Being found in forms at the level of the hydrogen atom. Among the *generators*, though, the diameter of every quantity of actual Being is one-half that of the diameter of the smallest quantity of actual Being ever found in the *forms*. As a result, among the generators, the diameter of the smallest quantity of actual Being is 2^{-129} rather than 2^{-128} times the diameter of the actual Being found in forms at the level of the hydrogen atom.

involved when a piggyback form changes from primary act to primary act. Basically, it means a beating heart, because, as with human hearts, piggyback forms alternately expand and contract. First, each piggyback form performs the primary act standard to its OD, which is to say it converts to actual Being that percentage of Z associated with its current OD. Next, it multiplies that percentage of Z by a maximum of 2. Next, it goes back to the standard percentage, and so on for as long as it remains at a given OD.

K. NATIVE OD & ITS IMPORTANCE:

Here's yet another point we must first make: Esoptrics says every form is created *native* to a particular OD. In this the 9th epoch of the universe, only 1 form is created native to $OD2^{256}$; 6 are created native to each of the ontological distances from 2^{255} thru and including $2^{256}-1$; 6^2 are created native to each of the ontological distances from 2^{254} thru and including $2^{255}-1$; 6^3 are created native to each of the ontological distances from 2^{253} thru and including $2^{254}-1$; etc.; until, at OD1, there are 6^{256} (*i.e.:* 1.61×10^{199}) piggyback forms native to it.

For Esoptrics, the native OD of a form is a very important factor. In the first place, as long as every form remains at its native OD, the universe is in an *un*accelerated condition. The instant a form takes on an OD different from its native OD, the universe is in an *accelerated* condition. In the second place, the native OD of a form determines how it shall always rotate around 2 of its 3 axes (How it shall rotate around its third axis is determined by its current OD—one which may or may not be its native OD.). In the third place, the native OD of a form determines to what extent a form shall expand or contract in each of its changes from act to act.

L. WHY TIME SLOWS AS VELOCITY INCREASES:

You may have noted I've so far disregarded the issue of how often *generators* change from primary act to primary act. Even now, I'm not going to say much. Hopefully, though, this little shall be enough.

Whenever the OD of a piggyback form is *in*creased, the frequency of its change from primary act to primary act *de*creases. As you might expect, though, in the logic of a *mirror*, the reverse is true of the piggyback form's carrying generator; and so, the frequency of the generator's change from primary act to primary act *in*creases whenever the frequency of its piggyback form *de*creases.

It should be rather obvious what that means: Its increase to a higher current OD means the form shall change less frequently from primary act to primary act, and that drop in frequency of change shall necessarily mean the form shall move more slowly thru our world's kind of time, which is to say the kind of *discontinuous* time which measures change. In the opposite direction, its increase to a higher frequency of change means the piggyback form's carrying generator shall change more frequently from primary act to primary act, and that rise in frequency of change shall necessarily mean the carrying generator shall move more rapidly thru our kind of time. That's Esoptrics' way of agreeing with Einstein on this point.

M. THE RELATIVE DURATION OF THE PRIMARY ACTS OF A FORM:

While performing the same one primary act, no given form is *involved* in our world's only kind of time—namely: the kind which measures change. Nevertheless, because forms change from primary act to primary act at different rates, it is possible to describe the duration of one form's primary acts by comparing that duration to the duration of some other form's individual primary acts. Clearly, though, that only tells us what the *relative* duration of this-or-that primary act is.

The best way to go about doing that is to make piggyback forms at OD1 the absolute standard. That's because:

FOR ALL THE UNIVERSE'S ULTIMATES THROUGHOUT HISTORY AND EVERYWHERE, THE FASTEST RATE AT WHICH CHANGE FROM PRIMARY ACT TO PRIMARY ACT TAKES PLACE IS FOREVER THAT WHICH TAKES PLACE AT OD1.

What is the duration of every primary act at OD1? It is impossible to say what it is *absolutely*; and so, time is never the *absolute* measure of the duration of any primary act on the part of an ultimate. All we can ever say is that the duration of this-or-that *higher* ultimate's every primary act is such-and-such relative to the duration of every primary act on the part of an ultimate at OD 1.

As long as the universe is in a non-accelerated condition, we can set up a formula expressing the duration of every primary act on the part of any and every form whose native and current OD is greater than 1. That formula tells us this:

FOR EVERY FORM WHOSE NATIVE OD IS GREATER THAN ONE: WHEREVER A GIVEN FORM IS STILL AT ITS NATIVE OD, THE DURATION OF EVERY PRIMARY ACT ON THE PART OF THAT GIVEN FORM EQUALS THE DURATION OF A PRIMARY ACT AT OD1 TIMES THE SQUARE OF THE GIVEN FORM'S NATIVE OD.[1]

There is a kind of corollary which goes along with that. How so?! As I pointed out earlier, Esoptrics says every from changes from primary act to primary act by pulsing outward and then contracting inward.[2] That means we can speak of change on the part of piggyback forms as having a cycle composed of two phases, which is to say composed of two changes. That allows us to say this:

FOR EVERY FORM WHOSE NATIVE OD IS GREATER THAN ONE: WHER-

---

[1] Remember what was said on page 86 about the correlation between OD and the size of the form's actuality. That being so, we can say this: The duration of a form's act equals the square of its actuality's diameter relative to that of forms native to and currently at OD1. It's basically the same as saying the duration *increases* and the frequency of change *decreases* as the square of the distance.

[2] This winds up being the way Esoptrics explains the elliptical shape of the planetary orbits.

EVER A GIVEN FORM IS STILL AT ITS NATIVE OD, THE DURATION OF EVERY <u>CYCLE</u> ON THE PART OF THAT GIVEN FORM EQUALS THE DURATION OF A PRIMARY ACT AT OD1 TIMES TWO TIMES THE SQUARE OF THE GIVEN FORM'S NATIVE OD.

Let's consider some examples of what that means. If a given form is currently at its native OD of 2^{128}, then we can say this: The square of 2^{128} is 2^{256}; and so, for that given form, the duration of each of its primary <u>acts</u> is 2^{256} times that of primary acts on the part of a form native to, and currently at, OD1. Naturally, that then allows us to say this: For the given form, the duration of each of its *cycles* is 2^{257} times that of primary acts on the part of a form native to and currently at OD1.

Obviously, that then allows us to say that the duration of every primary act—on the part of a form currently at its native OD1—is 2^{-256} (Notice the minus sign.) times that of primary acts on the part of a form native to, and currently at, OD 2^{128}. So what?! As any one can rather readily see, that's a case of arguing in a circle; and so, in the final analysis, we're not saying anything about the *absolute* duration of any form's primary acts. Speaking in the terms of *absolutes*, time is simply never *really* a measure of the duration of any primary act whatsoever on the part of an ultimate constituent of the universe. At least, that's the way it is with regard to the only kind of time present in the universe and, thereby, to our acts of awareness—namely: that discontinuous time which measures change.

One more example! Only one macroscopic ultimate is native to OD2^{256}. Throughout this the ninth epoch of the universe, it is the upper (Say "outermost", if you prefer.) boundary of the universe. For that reason, I call this particular macroscopic ultimate the "form of the universe" and the "tenth categorical form".[1] The square of 2^{256} is 2^{512}. From that, we deduce that the duration of each primary act on the part of this form is 2^{512} times the duration of each primary act on the part of forms native to, and currently at, OD1. Double that to calculate the duration of each of its cycles.

In an *accelerated* universe, the duration of a given form's primary acts equals its current times its native OD. For example, take a form native to OD2^{128}, and accelerate it to OD2^{256}-1.[2] The duration of each of its primary acts will in that case be: $2^{128}(2^{256}-1)K = (2^{384}-2^{128})K$, where K equals the duration of the primary acts of a form native to, and currently at, OD1.

According to Esoptrics, the duration of this the 9th epoch of the universe is 2^{385} K. If that be correct, then, in the course of the 9th epoch, every form native to OD2^{128} and accelerated to OD2^{256}-1, shall perform 2 acts, which is to say 1 cycle.

As we shall eventually see, that means every such form shall rotate exactly 360° around what Esoptrics calls its "tertiary" axis. Inevitably, that means this: For

---

[1] Shades of St. Thomas Aquinas et al who maintained that the "tenth crystalline sphere" was the outermost boundary of the created world! For Esoptrics, the ten *categorical* forms are those native to OD2^0 (*i.e.:* OD1), OD2^1, OD2^2, OD2^4, OD2^8, OD2^{16}, OD2^{32}, OD2^{64}, OD2^{128}, and OD2^{256}.

[2] Esoptrics says this produces a maximally massive particle in which 2^{256} –1 piggyback forms and their carrying generators are all concentric to one another. It's Esoptrics' version of a black hole.

Esoptrics, the entire universe rotates, but does it so slowly, it remains to be seen if we could ever observe it, though we might detect something which *implies* it.

N. THE CORRELATION BETWEEN SIZE & RATE OF CHANGE:

In the above principles regarding rates of change, Esoptrics is telling us something very important—namely: The smaller the targets we investigate, the more frequently shall they change from primary act to primary act. Esoptrics thus predicts that the molecular, atomic, and sub-atomic worlds (*i.e.:* the worlds we at least attempt to examine with our *micro*scopes) should be far more active than the worlds we examine with our senses and our *tele*scopes. It's a prediction which scientists all around the globe verify on a daily basis. What's even more astounding is that Esoptrics exhaustively explains exactly *why* the *micro*scopic world is more active than the *macro*scopic one. What other theory does as much?[1]

My youngest brother appeals to atomic clocks as evidence that time is continuous. He is sadly mistaken. Atomic clocks tell us nothing save that Esoptrics is right on the mark when it says: Diminish the size, and you increase the frequency of change and, consequently, the rate of involvement in time. When it comes to the nature of time and its ability to be relative, Esoptrics gives an answer incalculably simpler, clearer, and more thorough and satisfying than any other theory comes close to giving.[2]

O. WHY SOME CAN'T GRASP THE NOTION OF TIMELESS MOMENTS OF DURATION:

---

[1] Esoptrics thus also says this: As long as forms are detained within the confines of an atom, they have the potential to change from primary act to primary act at rates so high, they thus also have the potential to produce secondary actions in which there are velocities of as much as 2^{128} times the speed of light. Needless to say, if velocities that high are confined to an area as small as maybe 10^{-47} centimeters, they are undetectable save perhaps as incredibly high energy levels.

[2] **NOTE OF DECEMBER 22, 2008:** Most certainly, atomic clocks do not confirm by observation that time is continuous, which is to say they do not empower us to observe for a fact infinitesimal segments of time. All they do, at best, is *imply* that time is continuous. And how do they do so? Is it because they themselves imply it by their very nature, or is it a case of the observer's psychological needs forcing the clocks to imply what inference the observer *has* to see implied? But, whether the inference of infinite divisibility is implied by the clocks on their own or implied only to eyes with ulterior motives, an inference is an inference and, as such, requires corroboration. Indeed, in the face of Esoptrics' explanations and the arguments of Zeno et al, the inference of infinite divisibility requires an ocean of corroborating evidences more voluminous than the Pacific. The trouble here is this: As with virtually all carnal minded scientists, my brother cannot tell the difference between what's actually observed and what is merely an inference implied—genuinely or fraudulently—by what's observed. As a result, he cannot here perceive a need for any corroborating evidence at all, let alone the need for an *ocean* of corroborating evidence. In the face of Esoptrics and Zeno et al, Science must either present us some kind of way to observe infinitesimal segments of time, or it must face up to the fact that continuous time—like continuous space, spatial bodies, and locomotion—is a sacred cow in its death throes.

Why do others find it impossible to grasp the above division between time-less moments of duration and durationless moments of time? Well, as one can probably observe for one's self, we do not *experience* the duration of any of the timeless moments of duration. As far as we can *experience*, neither we ourselves nor anything around us ever stands still for even the tiniest fraction of a second. Contrary to what everyone but myself *apparently* thinks, that proves nothing no matter how strongly experience seems to *imply* continuous time. How can I say that?

On the face of it, where there is no change, there is neither any *involvement* in time nor any **_experience_** of the only kind of time we know; and so, it's not possible for us to *experience* the *duration* of our individual acts of consciousness. We, then, necessarily experience each individual act of consciousness as durationless— just as, if you sleep soundly, several hours go by in a flash. More accurately, because of that *apparent* lack of duration, experience finds consciousness a single act ceaselessly changing and, consequently, ceaselessly involved in time. That, how-ever, is an experience which—according to Esoptrics—is thoroughly mistaken.

P. IS EXPERIENCE, THEN, THE WORK OF A DECEIFTUL GOD?

What, then?! Is not consciousness itself a monstrously crippling deception necessarily proclaiming its Creator an incredibly deceitful God? And what's the point of this "monstrously crippling deception"?

Monstrously crippling deception???!!! If so, how is it Zeno of Elea saw thru it 2,400 years ago? If so, how have *I* managed to see thru it? And bear in mind that I have no *formal* education beyond high school and first saw thru this "deception" over 50 years ago. Yes, according to those who tested me on a few occasions, I have an I. Q. in the genius range; but, it's in the *lower* part of that range. How, then, does a high-school graduate in his early twenties and only slightly a genius overcome a "monstrously crippling deception", unless the "monstrously crippling deception" isn't all it's cracked up to be?

> *Hugh of St. Victor*, in the hope of affiliating Mysticism and Scho-lasticism, collected and arranged systematically the scattered thoughts of St. Bernard favorable to his purpose. With him, the underlying prin-ciple of religious science was that one's knowledge of truth is exactly adequate to his interior dispositions. (*Tantum de veritate quisque protest* [Sic!] *videre, quantum ipse est*. [The quantity of truth one is able to see is directly proportional to what much one is – ENH.]) The means of ar-riving at perfect science is contemplation, which was lost through origi-nal sin, yet can be recovered by supernatural aids. This fixing of the mental vision on things eternal is what is understood by contemplation in the strict sense. When, on the other hand, the faculties are engaged in the consideration of the objects that meet one in the visible world, the mental operation is called rational meditation.
> ——**REV. DR. JOHANN BAPTIST ALZOG** (1808 – 1878 A. D.): **_Man-ual Of Universal Church History_**, § 256; pgs. 763 & 764 of volume 2 of the 3 volume set translated by F. J. Pabisch & Rev. Thomas S. Byrne,

and published by Robert Clarke & Co., Cincinnati OH, 1874 [The "Sic!" note above adverts to the fact that, in Fr. Alzog's book, the spelling is indeed *protest* instead of *potest*.].

Well, I guess that's one way to explain it; but, that would imply I have somehow (thru divine aid?) overcome the effects of original sin more so than most do and, thereby, have achieved an extraordinary level of contemplation (*i.e.:* introspection) leading to an extraordinary level of what Fr. Alzog calls "religious science". I'll go no further down that path of egomania save to add this well known Latin axiom: "*Nemo Judex in causa sua.*"

As for the *point* (*i.e.:* purpose) of this "monstrously crippling deception", it's easy to explain it once you get at least a moderate amount of knowledge regarding the universe's <u>macro</u>scopic ultimates. What it boils down to is this: It's an incredibly *permissive* God's way of allowing as many different types of individuals as possible to inhabit the universe, and to do so in a way which allows them to interact conveniently *with their own kind*. What does that mean?

Q. THE MEANING OF
"TO INTERACT CONVENIENTLY WITH THEIR OWN KIND":

The answer to that shall require us to touch briefly on the issue of space. Perhaps you shall forgive me if I now *briefly* violate what I said earlier about limiting this letter to the issue of time.

UNIVERSE NUMBER	ONTOLOGICAL DISTANCES INCLUDED IN	
	THE MAJOR UNIVERSE	THE MINOR UNIVERSE
FIRST	1 thru 1 = 2^0 thru 2^0	1 thru 1 = 2^0 thru 2^0
SECOND	2 thru 3 = 2^1 thru $2^2 - 1$	2 thru 3 = 2^1 thru $2^2 - 1$
THIRD	4 thru 15 = 2^2 thru $2^4 - 1$	4 thru 7 = 2^2 thru $2^3 - 1$
FOURTH	16 thru 255 = 2^4 thru $2^8 - 1$	8 thru 15 = 2^3 thru $2^4 - 1$
FIFTH	256 thru 65,535 = 2^8 thru $2^{16} - 1$	16 thru 31 = 2^4 thru $2^5 - 1$
SIXTH	2^{16} thru $2^{32} - 1$	32 thru 63 = 2^5 thru $2^6 - 1$
SEVENTH	2^{32} thru $2^{64} - 1$	64 thru 127 = 2^6 thru $2^7 - 1$
EIGHTH	2^{64} thru $2^{128} - 1$	128 thru 255 = 2^7 thru $2^8 - 1$
NINTH	2^{128} thru $2^{256} - 1$	256 thru 511 = 2^8 thru $2^9 - 1$
TENTH	Not yet existent.[1]	512 thru 1,023 = 2^9 thru $2^{10} - 1$

CHART #2: LEVELS OF THE UNIVERSE
AT WHICH INTELLIGENT BEINGS CAN RESIDE

Esoptrics suggests that space itself is *four* dimensional. Upon its saying that, do not imagine that—for Esoptrics as for Einstein—"fourth dimension" means

---

[1] According to Esoptrics, the tenth major universe shall come into existence as, at OD1, there are one more than 2^{385} (*i.e.:* 7.88 x 10^{115}) successive changes from primary act to primary act.

time. For Esoptrics, this fourth dimension is ontological distance, because the quantity of Being actualized in each act of a piggyback form increases as its location in the fourth dimension is logically raised. According to Esoptrics, there are 2^{256} (*i.e.:* 1.1579×10^{77}) levels (*i.e.:* ontological distances) in the universe's fourth dimension. Every piggyback form is created *native* to one of these ontological distances and forever exhibits two characteristics which forever identify the OD to which it is *native*. I'll not discuss them here.

The way piggyback forms and OD's work, it's possible to have a universe (*i.e.:* "multi-verse" if you prefer) containing what we can call 255 different *minor* universes within 9 different *major* ones. The difference between one and the other is the range of OD's included. The chart on the prior page spells out that difference. You may quickly notice that, in the right-hand column, one is progressing by successive *powers* of 2 and, in the middle column, progressing by successive *squares* of the powers of 2. Needless to say, I'm not about to burden you with minor universes beyond the tenth one. Besides, you can probably figure them out for yourself.

I should perhaps point out in passing that the prior page's chart holds true only where there is no *acceleration*, which is to say no case of a macroscopic ultimate changing from its native OD to a higher one. Where no acceleration has yet taken place, a macroscopic ultimate dwelling in the 128th *minor* universe within the ninth *major* universe, would experience occupants of the eighth, seventh, sixth, fifth, fourth, third, second, and first major universes as being *micro*scopic in nature and would experience all co-occupants of the ninth *major* universe as *macro*scopic in nature. How different the view for occupants of the first major universe![1]

Contrary to what Science imagines as a result of its obsession with *micro*scopic ultimates (an obsession Leibniz shared with them), every act of consciousness is an act on the part of a *macro*scopic ultimate. But, every *macro*scopic ultimate dwells within some one of the minor universes within some one of the major universes; and so, to interact with its *own kind*, it must interact with those other macroscopic ultimates sharing the universe in which it dwells. How shall it do that?

It's rate of change from primary act to primary act must approximate that of the other macroscopic ultimates sharing the universe in which it dwells. That's just another way of saying that its rate of involvement in time must approximate that of the macroscopic ultimates *of its own kind*. Again, that's just another way of saying its *own* OD (as opposed to the OD's of the piggyback forms governing its cells, molecules, atoms, and sub-atomic particles) must remain within the range of the OD's contained within its minor universe.[2] The *convenience* of that interaction of like with

---

[1] Esoptrics thus tells us that, for occupants of the ninth major universe, their microscopic world includes 127 minor universes. One of these makes room for electrons (They're the product of piggyback forms native to ontological distance 1, and, early on, my theory told me much about their behavior: how they should normally radiate from the atom at the speed of light, and how they should move up and down in ontological distance by ingesting and disgorging what Science apparently calls neutrinos.). The other 126 make room for 126 different kinds of atomic nuclei. Disgustingly, I've never been able to figure out how the latter work. [As of 2008, that's no longer true.]

[2] In death, this becomes impossible without divine assistance; and so, the macroscopic ultimate, which each of us is, descends to an ontological distance of perhaps as low as 1. That would put us at

like is then made possible by the fact that—being oblivious of the duration of their individual acts and aware only of their rate of involvement in time—they all experience themselves and their kind as *continuously* involved in interacting with one another. Would any sane individual want it any other way?

R. THE RELATIONSHIP BETWEEN TIME & HISTORY:

Let's now try to summarize to some extent what's been said here. To that end, let's turn to what you wrote in the last part of your e-mail:

> Incidentally, though I don't care very much whether space is real or not I do care whether time is real. things [Sic! No capital "T"!] have causal and logical relations to each other --the premises come before the conclusion, even in logic, and so if time is not real you have difficulty making sense of causal and logical relations. I think this is a pretty high price to pay for the claim that basic entities are a-temporal [Sic! No period!] Don't you think that first Adam had to sin, and THEN Jesus came to redeem the sin? If time is not real, Christianity makes no sense.[1]

Your words above *suggest* to me that you are at least unconsciously equating time with history. For you, it *seems*, "where in history", "where in time", "at what point in history", and "at what point in time", all signify basically the same factor. If so, then you are sadly mistaken—according to Esoptrics, that is. How so?!

Esoptrics says that—for all forms everywhere in the 9th epoch—"where in history" signifies the same point; but, "where in time" does not. How can that possibly be? Let's answer with some examples.

For all forms everywhere, halfway into the history of the 9th epoch is the same, one point in that history. The same can be said of one-quarter into that history, or one-sixteenth into it, or 1/32, or 1/64, and so forth. But, look what half-way into that *history* means to forms when it comes to *time*: For a form native to, and still at, $OD2^{192}$, it means one brief moment into time; for a form native to, and still at, $OD2^{128}$, it means 2^{128} brief moments into time; and for a form native to, and still at, OD1, it means 2^{384} brief moments into time. For each of those forms, the half-way point in the history of the 9th epoch is the same point; **BUT**: For an $OD2^{192}$ form, that half-way point is reached in "the blink of an eye"; for an $OD2^{128}$ form, that half-way point is reached only after more eye blinks than you or I can make in roughly 10^{32} of our years; and, for an OD1 form, that half-way point is reached only

---

the center of the Earth in a sense far more accurate than the notion that the center of the Earth is the core of the planet. For, it puts us in what is in the center of every *smallest piece* of earth.

[1] **NOTE OF AUGUST 11, 2009:** I never said *time* is not real. I ever said *continuous* time is not real. The issue is not whether or not time is real; it's *what kind* of time is real. How is that not even a highly reputable Prof./Dr. of Philosophy can follow me when I make so simple a distinction? To accuse me of saying either that time is not real or that basic realities are a-temporal is more outrageously blind than the human mind can comprehend or the human tongue describe.

after more eye blinks than you or I can make in over 10^{100} of our years.[1]

For all forms always and everywhere, the *order* in every sequence of events is the same. All that's different is *how much time* it takes for that sequence to run its course. For all forms always and everywhere, Adam sinned first and then Jesus came to redeem; and so, for all forms everywhere, the former event preceded the latter in *history*; **BUT**: For forms such as ours, the former preceded the latter by only a few millennia; for other forms, the former preceded the latter by only a tiny fraction of the time it takes to blink an eye; and, for still other forms, the former preceded the latter by the amount of time it takes to blink an eye trillions of trillions of trillions of trillions of trillions of trillions of trillions of times. What precedes what in *history* is the same for all forms everywhere; but, if you ask how much *time* elapsed between what and what in *history*, the answer is purely relative to the OD of the form you're interrogating. It's because the measurer's attempts to measure time are always greatly impacted by the rate at which the measurer changes from primary act to primary act. Alas! As simple a principle as it is, it's one a Science infected world can never grasp as long as its fixation upon *micro*scopic realities closes its mind to thinking in the terms of *macro*scopic ones.

Hopefully, you can now see how wrong it is to say Esoptrics presents us with "basic realities" which "do not exist in time but are outside of time." If you do not, what can I say? Well, maybe it shall help to close with this: We cannot *sanely* equate *changeless duration* with *durationless change*. They are so totally opposites, no mentally competent person could possibly fail to see in a flash that they cannot be lumped together as one and the same *species*. Yes, we might be able to classify them as two mutually opposed species of the same *genus*. For example, they might both be called a type of *moment*. Still, to lump them together as one and the same *species* would be ludicrous. By the same token, we cannot *sanely* equate what *measures* changeless duration with what *measures* durationless change. They are so totally opposites, no mentally competent person could possibly fail to see in a flash that they cannot be lumped together as one and the same *species*. Yes, we might be able to classify them as two mutually opposed species of the same *genus*; but, the instant we call that *genus* "time" and try to say the term "time" applies to each of them equally, we merely resort to buffoonery.

Finally, I ask you this: Since—as far as we can observe—we never to any extent observe what remains changeless for even the tiniest fraction of a second, how can we *possibly* know anything whatsoever about what measures the duration of

---

[1] **NOTE OF DECEMBER 22, 2008:** Remember what we said earlier: For every form native to, and currently at, $OD2^{192}$, the relative duration of each of its acts is $(2^{192})^2 K = 2^{384} K$. But, the relative duration of the 9th epoch—from its inception to the inception of the 10^{10} epoch—is $2^{385} K$, where K = the rate at which every form native to and still at OD1 changes from primary act to primary act. Therefore, in the course of the 9th epoch's history, a form native to, and currently at, $OD2^{192}$ would perform only two primary acts, and that necessarily means it reaches the mid point of the 9th epoch's history in what—according to its kind of involvement in time—is a single moment of time. By the same token, for a form native to, and still at, 2^{128}, the relative duration of each of its acts is $(2^{128})^2 K = 2^{256} K$. Naturally, for it to reach the halfway point of the 9th epoch's history at $2^{384} K$, it must change from primary act to primary act 2^{128} times and, therefore, reaches that halfway point in 2^{128} moments of time according to its kind of involvement in time. Why is all of this virtually impossible for anyone but myself to follow?

what endures in a changeless state?

That's enough on time. In my next letter, I will tackle the issue of space. For now, let me say no more on that issue than this: It is one thing to say space is not real; it is quite another to say that a kind of space—a kind which is extended by its very nature (*i.e.:* in, of, and by virtue of, its own self rather than by virtue of some exterior influence constantly maintaining that extension)—is not real. *Self*-extended space and spatial realities are a farce. The issue is not: inside vs. outside space; rather, it's: inside *physically* extended space vs. inside *logically* extended space.

Finally, here's one of the quotes dear to my heart:

First, we must ask if—as most philosophers of yore concluded—some kind of infinity is found in material bodies. Or, is that impossible? Whichever way you answer that question, it's by no means <u>un</u>important. On the contrary, it's *all important* to our quest for truth. That's because this particular issue has almost always been the chief cause of the feuds among those who have written about nature. That's the way it *has* been and *must* be. For, no matter how little you stray from the truth at the start, it winds up a thousand times worse down the line.
——**ARISTOTLE: *On The Heavens;*** 271^b, 2-9. My rather free translation – ENH. Compare with page 326 of volume 8 of **Great Books Of The Western World** as published by William Benton, Chicago, 1952.

As a child in the first or second grade, I answered yes it is impossible. How it comforted me around 15 years later when I learned that at least Zeno of Elea and Leibniz agreed with me! I may not have been crazy after all.

Until the next time, then, most generous and patient Sir:

May the Lord bless you and keep you,
And lift up His Countenance upon you.
Yea, may He make His Face to shine upon you,
And give you Peace.

In Christ Jesus,
    Thru Mary
        With Joseph,
A Brother To All:

EDWARD N. HAAS

שָׁלוֹם

Completion Date Of This Version: Friday, September 07, 2007

POSTSCRIPT OF AUGUST 21, 2009: Throughout what's about to follow, please bear in mind that, for Esoptrics, "time" means discontinuous time. As, hopefully my reader has seen, that means the kind of time which measures rate of change from primary act to primary act.

In retrospect, this letter on time seems lacking. Maybe I can remedy that with the following distinction between *subjective* and *objective* time.

Objective time tells us how many alphakronons it takes for *any* given event or sequence of events to run its course. When it does that, it can be called "objectively relative time", since it alone describes relative duration as perfectly as it can be described—namely: in the terms of the number of alphakronons involved. It's also able to tell us at *exactly* what *point* any given event or sequence of events started and/or stopped in the terms of number of alphakronons since the beginning of the universe's 9th epoch. When it does that, it can be called "*historically absolute* time", because it alone describes location in history as perfectly as it can be described—namely: in the terms of the number of alphakronons involved. It can also be called "universal time", because it alone applies throughout the universe, as long as the issue is not what's below 10^{-96} seconds.

Subjective time is time as it applies to some *one particular* ultimate and measures the rate at which it and it alone changes from primary act to primary act. Subjective time applies to its individual subject only in the durationless instant in which that subject is changing *from* what—in that durationless instant—is a *past* internal condition *to* what—in that durationless instant—is a *future* internal condition. Subjective time is either experienced time or non-experienced time. The former applies to an individual subject able to be *aware* of subjective time; the latter applies to an individual subject *not* able to be aware of subjective time. The former can be called—to use Einstein's term—"I-time"; the latter can be called "it-time".

I-time can measure how long it takes an event or sequence of events to run its course according to some conscious individual's experience of time. According to Esoptrics, that means, for most, how long an event or sequence of events takes to run its course according to the number of the observer's acts of consciousness, which, for us, means multiples and fractions of roughly .03 seconds. When it does that, I-time can be called "subjectively relative time", because it thus describes relative duration imperfectly. I-time can measure where in history which event occurred. When it does that, I-time can be called "*historically relative* time", because it thus describes that place in history imperfectly. It can also be called "local time", because it applies only to those who share the same rate of change from primary act to primary act. I-time is the same as absolute time, if—as in the case of angels and, perhaps, some of those in hell or Purgatory—it pertains to a conscious ultimate changing from primary act to primary act at the maximum rate.

It-time measures how long it takes an event or sequence of events to run its course relative to the rate at which some mindless ultimate changes from primary act to primary act. It-time is the same as absolute time, if it pertains to a mindless ultimate changing from primary act to primary act at the maximum rate.

Objective time measures only: (1) objectively relative duration regardless of whether or not it involves change from past to future, and (2) what has what absolute location in the history of the universe's 9th epoch. It is subjective time alone which measures *individual change from past to future*. For that reason, objective time is what time *really is* as long as we are concerned either with objectively relative duration or with absolute location in history. Subjective time is what time *really is* as long as we are focused either upon our own motion from past to future or subjectively relative duration or relative location in history. Without Esoptrics, subjective time is, for us, the only kind of time which *really is* time.

Because they are very distinct from one another, subjective and objective time are *outside* of one another. For all of that, they ever remain *alongside* of one another. They are two *separate* but *concomitant* streams neither of which excludes the other from the whole picture. They are thus not outside of one another in the sense that neither has any kind of connection whatsoever with the other. On the contrary, they are so connected that—however much subjective time is outside of objective time—it has no effect upon what comes before what according to objective time. Whether for philosopher or scientist, the trick is to understand exactly how these two kinds of discontinuous time are connected.

See, then, how our kind can transcend Prof. Hawking and his fellow Positivists and find out what time *really is*: Because *Science* can neither observe nor in any way confirm what changes every 10^{-96} seconds, *Science* cannot tell us what time *really is*; but, *Esoptrics* certainly can, and it can do so because *Philosophy*—at least in the hands of Edward N. Haas (Oh delicious ego trip!)—is not nearly as dead as Wittgenstein alleged.

Lack of attention to externals lowers one's ability to achieve a *thorough* grasp of what we all already accept as fact. That's why we see this: Some live their whole lives keenly focusing upon what the external world presents to us. _They_ beget theories founded on principles ever able to grow in breadth and harmony. Others, having made few external observations, quickly limit themselves to abstruse argumentation. _They_ beget only hasty dogmatism. The former is the *scientific* method of inquiry; the latter is the *dialectical* one.
——**ARISTOTLE: *On Generation And Corruption***, Book I: Chap. 2; 316ª: 5-3. My rather free translation - ENH. Compare page 411 of volume 8 of **Great Books Of The Western World** (Curiously, this is the same complaint science has been hurling at Aristotle for at least the last 400 years. And is it not ironical? Like Aristotle, the scientists—men who "live their whole lives keenly focusing upon what the *external* world presents" to them—cannot discover what time *really is*. Ah, but! The philosopher—a man who, "having made few *external* observations, quickly" limits himself to what his own *internal* world presents to him (and this includes a lot of what Science would call "abstruse argumentation")—can indeed tell us what time *really is*. Who now is the "they" who "beget only hasty dogmatism"?)

Document #5:

LETTER BEGUN JULY 7, 2007

A: INTRODUCTORY:

Prof. D. . . ., Chair
Dept. Of Philosophy
University Of . . . ., NY

Greetings, again, Prof. D, my dear, kind and generous Sir:

Did you receive and read my June 8th letter regarding time and, especially, primary vs. secondary acts? I am assuming that you did; and so, if, in fact, you did not, then there shall be occasions when you'll not be able to follow me, because you'll not have what preparatory remarks I'm assuming you have. Of course, my nephew, the emergency room surgeon, would probably say: "Don't worry, Uncle Edward. He won't be able to follow you no matter what, because nobody can ever understand anything you write." He's very honest with his uncle that way.

May I remind you that I choose to refer to my cosmological theory as "Esoptrics", and that is short for "The Divine Algebraic Logic Of The Mirror". It's also short for "Haasian Cosmology". As I said in the prior letter, if I say "Haasian Cosmology", some accuse me of being conceited. So what?! To each his own!

B: ESOPTRICS VS. LEIBNIZ ON TIME:

Now, let's return to your e-mail of Monday, June 4, 2007, 17:52:05. Immediately after the first sentence, you wrote:

> Your basic hypothesis seems to be that basic realities do not exist in space and time but are outside of space and time. This is the view of Leibniz and other members of the 17th c. rationalists school: Leibniz held that the basic entities, or monads, are unextended centers of energy, and that space and time are phenomenal, pertaining to appearance, not real.

Believe me: I am reasonably well aware of what Leibniz [Some spell it Leibnitz] said, and I can assure you there is no comparison between what he said and what Esoptrics says. In my book *Introspective Cosmology II*, Chapter Thirty-Six is entitled: "Haas Vs. Leibnitz". It's an 8 page attempt to lay out the marked differences between us. I'll not here go into *every* one of those differences.

In my June 8 letter to you, I went into the difference between Esoptrics and Leibniz regarding time, which is to say between Esoptrics and those who say time is unreal and that "basic realities" (your term) are outside of time. Hopefully, you read that letter and followed it well enough to be firmly convinced now of this: Contrary to what Leibniz said (or truly intended to say, or is commonly interpreted to have said), Esoptrics says basic realities are by no means outside of time; rather, they are merely *intermittently* outside of time *provided* "time" means the discontinuous measure of change. Regardless of what Leibniz said (or truly intended to say, or is commonly interpreted to have said), Esoptrics does *not* by any means say that *time* is merely "phenomenal, pertaining to appearance, not real" (your words). What Esoptrics *does* say about time is that <u>continuous</u> time is merely "phenomenal, pertaining to appearance, not real." Time is the measure of *change*; but, ultimates change intermittently at various rates; and so, what measures those rates must itself be intermittent and, so, cannot be continuous. What measures change from past to future cannot also measure the *duration* of what is not so changing.

I repeat: For Esoptrics, the issue is by no means: whether or not *time* is real; rather, the two issues are: (1) whether or not <u>continuous</u> time and change are—as they *appear* to be—real; and (2) whether or not the same one yardstick (***viz.:*** time) can measure both the motion of what's changing from a past to a future condition and the duration of that which is not performing such a change. Again, for Esoptrics, the issue is by no means: whether or not basic realities are outside of *time*; rather, the issue is: whether or not basic realities are outside of <u>continuous</u> time and change. If, as Esoptrics contends, *continuous* time and change are merely the figments of misguided minds, then, yes, basic realities are outside of *continuous* time and change for the same reason *every* real thing is outside of Alice's Wonderland—namely: Alice's Wonderland is a mere fairy tale and a figment of the human imagination, and everything real is outside of fairy tales and mere figments.

But, didn't I say enough about time in my June 8 letter? I am reasonably sure I did; and so, let us pass on to the issue of space.

C: ESOPTRICS VS. LEIBNIZ ON SPACE:

Contrary to what Leibniz said (or truly intended to say, or is commonly interpreted to have said), Esoptrics is far from saying either: (1) that basic realities do not exist in space and, rather, are outside of space; or (2) that the basic entities are unextended; or (3) that space is phenomenal, which is to say pertaining to appearance only; or (4) that space is not real. On the contrary, Esoptrics says:

FIRST: Basic realities do not exist in *physically extended* space but, rather, are outside of *physically extended* space for the same reason every real thing is outside of Alice's Wonderland—namely: Alice's Wonderland is a mere fairy tale and a figment of the human imagination, and everything real is outside of fairy tales and mere figments. Since, for Esoptrics, basic realities are real—but *physically extended* space a fairy tale and a mere figment of misguided minds—the universe's ultimate constituents are most certainly outside of *physically* extended space.

SECOND: Basic entities are by no means *non-extended*. They are *not physically* extended and, instead, are *logically* extended. By "physically extended", I mean what's extended by its very nature, which is to say: in and of its own self. To put it another way: The *physically* extended needs no explanation of how it manages to be extended, because that's just the way it is—**PERIOD**. In short, what I mean by the *physically* extended is the "*res extensa*" so dear to the heart of Descartes[1] and perhaps 99.999999% of all the men, women, and children on the street regardless of I. Q. or level of education.

THIRD: It is *physically* extended space which is phenomenal. It is *physically* extended space which pertains to appearance only. Whether space, spatial objects, or what have you, any and every instance of *physical* extendedness is merely a figment of the misguided mind of the individual surmising he or she is gazing upon an instance of *physical* extendedness. It is *logical* extendedness which is real, and *physical* extendedness is merely the self-contradictory gibberish of a blundering mind still too dark either to comprehend the nature of *logical* extendedness or to perceive that, with *physical* extendedness, "we entangle ourselves in the labyrinth of the continuum and in inexplicable contradictions."[2]

FOURTH: It is *physically* extended space which is not real. Only *logically* extended space is real.

For Esoptrics, the issue is not: whether or not space is real; rather, it's: whether nor not *physically* extended space is real. Is *physically* extended space real and *logically* extended space a fairy tale, or vice versa? For Esoptrics, the issue is not: Are basic realities extended, or are they non-extended? Instead, the issue is: Are basic realities *physically* extended, or are they *logically* extended? Is physical extendedness real and logical extendedness a fairy tale, or vice versa? What is so difficult about that? To me, it's as simple as 2 + 2 = 4; and yet, though I've been saying it for 45 years, it's seems I've not yet found a single, solitary other human being able to understand what I'm saying. All I get, on one hand, is Leibniz thrown in my face every time I open my mouth and, on the other, people who, as does my nephew, at least *effectively* tell me: "I'm an emergency room surgeon and about as highly intelligent and educated as anyone can be; and yet, I can't understand a word you write." Are my thoughts or ways of expressing them *that* unintelligible?

How is it possible to mistake what I'm saying for what Leibniz said? Where, in Leibniz (or, for that matter, in any "other members of the 17th c. rationalists school"), is there even so little as the slightest hint of a distinction between physical and logical extendedness? Where is there anything but space vs. no space and extendedness vs. no extendedness? And that's the same as what Esoptrics is saying???!!! Give me a break!!!

One might perhaps object: "Well, we might be able to see a difference be-

---

[1] See the upper half of the right-hand column of page 226 in the second edition of ***The Cambridge Dictionary Of Philosophy*** as published by the Cambridge University Press, New York, 1999.

[2] Leibniz as quoted by Prof. Robert C. Sleigh, Univ. of Mass., Amherst, in the middle of the left-hand column of page 492 of his article "Leibniz" in the second edition of ***The Cambridge Dictionary Of Philosophy*** as published by the Cambridge University Press, New York, 1999.

tween you and Leibniz if you were to give us a clear cut, concrete description of what logical extendedness means and how it's possible." I've already done that in other writings of mine and without getting thru to anyone, it seems. Still, as the saying goes, hope springs eternal in the human breast. Why not, then, try it again? Very well! Off to it!

D: LEIBNIZ'S MONADS VS. ESOPTRICS' HEBDOMADS:

Since you throw Leibniz and his monads at me, let's start with this bit about monads. What Esoptrics puts forth as basic realities can hardly be called "monads". How so?! I guess that depends upon what one means by the term "monad". For the moment, I suppose, it mainly depends upon what Leibniz meant by "monad". Looking at the word from the standpoint of its linguistic origins, one would be inclined to say: "'Monad' comes from '*monos*' which is the Greek word for 'alone' or 'one'. Therefore, 'monad' implies something which is very much a singularity, which is to say something in which there is but one and one factor alone."

Leibniz seems to have used the term in that sense and, thereby, to imply that each of his basic realities was wholly and entirely simple within itself. On the other hand, we read this:

> . . . he concluded that Descartes was mistaken in attempting to reduce matter to extension and its modifications. Leibniz concluded that each material substance must have a substantial form that accounts for its active force. These conclusions have to do with entities that Leibniz viewed as phenomenal. He drew analogous conclusions concerning the entities he regarded as ultimately real, i.e., the monads. Thus, although Leibniz held that each monad is absolutely simple, i.e., without parts, he also held that the matter-form distinction has an application to each created monad. In a letter to De Volder he wrote:

>> Therefore, I distinguish (1) the primitive entelechy or soul, (2) primary matter, i.e., primitive passive power, (3) monads completed from these two, (4) mass, i.e., second matter . . . in which innumerable subordinate monads come together, (5) the animal, i.e., corporeal substance, which a dominating monad makes into one machine.

——**ROBERT C. SLEIGH:** *Leibniz* as found in the middle of the right-hand column of page 492 of the second edition of ***The Cambridge Dictionary Of Philosophy*** as published by the Cambridge University Press, New York, 1999.

If, as Prof. Sleigh says, "Leibniz held that each monad is absolutely simple, i.e., without parts," then how does he then say each is "completed" from soul and primary matter? Isn't that *two* components? What, then, *did* Leibniz mean by "monad"? As for myself, I can only guess at the answer. Whatever that answer might be, though, one thing is certain: "Monad" could in no way apply to the basic realities put forth by Esoptrics. If, nevertheless, one is intent upon shoving Esoptrics into a Leibniz mold, then one must refer to Esoptrics' *basic realities* as "hebdomads" and its *ultimate constituents* as "quartuordecimads". Wait a while, and I'll eventually get around to the meaning of "quartuordecimads" and the difference between *"basic realities"* and *"ultimate constituents"*.

"Hebdomad" is a word you can find in virtually any dictionary, and it's generally defined as a group of seven. Because every week has seven days, every week is an example of a hebdomad. Needless to say, one cannot sanely equate a *monad* with a *hebdomad*, and that alone puts Leibniz and Esoptrics far apart.

In a further departure from Leibniz, who has only one kind of monad, Esoptrics has two radically different kinds of hebdomads—namely: generators and forms. Each is a hebdomad because, in each of them, one can *logically* distinguish between *seven* principles *logically* outside of one another and forever conjoined inextricably in every generator and every form. Those seven principles are: (1) the hebdomad's actuality, (2) its A potency, (3) its B potency, (4) its A' (*i.e.:* A prime) potency, (5) its C potency, (6) its B' (*i.e.:* B prime) potency, and (7) its C' (*i.e.:* C prime) potency. In every hebdomad, the A and B potencies are mutual opposites; the A' and C potencies are mutual opposites; and the B' and C' potencies are mutual opposites. That can be *graphically allegorized* so:

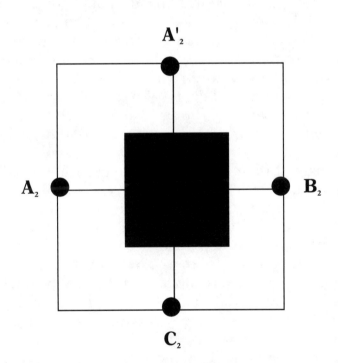

DRAWING #12: A FORM WHOSE ACTUALITY = .5 OF WHAT'S AVAILABLE; AND SO, EQUALS: .5A + .5A' + .5B + .5B' + .5C + .5C'

On this and the next several pages, the drawings are *geometrical allegorizations* of what are, in fact, purely logical principles; and so, these drawings are not *pictures* of what hebdomads *look like*. In these drawings, I am laboring under a severe handicap—namely: I cannot draw cubes, only squares. That leaves me unable to include the B' and C' potencies. You shall have to do that for yourself by *imagining* a cube whose third axis passes thru the center of the various squares.

Despite that handicap, I hope that, in looking at the 10 drawings, you'll get a preliminary idea regarding 3 important points: (1) the relationship between a form's actuality (*i.e.:* the solid black area in Drawing #12) and its six potencies (*i.e.:* the clear areas in #12); (2) the very different relationship between a generator's actuality (*i.e.:* the solid black areas in Drawings 13-21) and its six potencies (*i.e.:* the clear areas in 13-21); and consequently (3) the dramatic difference between a form and a generator, which is to say between the two types of basic realities called "hebdomads".

I dare say that, in Leibniz, you'll find nothing this graphic and detailed in his description of his basic realities. To compare Leibniz's ideas with ones this sophisticated is merely a joke.

The sub-scripts adjacent to the letters indicate that the dots next to the letters are each the second dot on that line. In Drawing #12, the actuality of the form has taken up the first dot on each line. In doing so, it has converted a kind of dead, passive, inactive, and sleeping form of reality into a kind of reality which is alive, aggressive, active, awake, and vital in a way we the finite can never imagine.

As for that "kind of dead, passive, inactive, and sleeping form of reality", we come face to face with it and experience its internal characteristics as often as we experience what we call empty space. In Esoptrics, you see, space is a kind of *reality*, and that's a far cry from people who, like Isaac Newton, deem space a globe of pure nothingness (Some deny he did.) which—despite being pure nothingness—nevertheless has length, width, and depth. What a miraculous kind of nothingness *that* nothingness is! And how *wrong* it thus is to accuse Esoptrics of saying space is not real! And how *strange* it is, when those who can readily accept *pure nothingness* as being real, cannot begin to accept logically extended space as real!

What is the origin of these 6 potencies? Be patient, and we'll get to that issue in due time. For now, suffice it to say they are the result of the heavenly sextets which, standing in infinity, serve as what teleology's devotees call the *final* cause. For all *forms* at all times everywhere, infinity is their absolutely final and chief goal; and so, they reach for it, and are "in potency" to it, every which way they can.

But, as we shall eventually see, there are logically 6 points of reference in infinity and, consequently, 6 ways to relate to infinity. As a result, every form, always, and everywhere, extends 6 arms to infinity (*i.e.:* 6 strings?), and these 6 are logically 3 sets each of 2 mutual opposites thereby producing what it's logical to signify by three axes each of which is perpendicular to the other two. As a result, every form is logically enveloped in a mantle of 6 potencies which—using the "Geometry speak" of Science—we can then signify by a cube approximately 10^{31} cm. (*i.e.:* 18 trillion light years) on each of its 6 sides and being composed of $(2^{257})^3$ tiny cubes each roughly 10^{-47} cm. on each of their 6 sides. In short, the 6 potencies are the result of an *extra-cosmic* influence (*i.e.:* an influence which, logically, is wholly outside the universe) which is incessantly affecting those sexpartite hebdomads I call "piggyback forms". Take away the extra-cosmic influence of that final cause, and, the *in-*

stant you do so, there is no such thing as extendedness of any kind whatsoever anywhere whatsoever. Without the *incessant* intervention of The Infinite's divine influence as *final* cause, there is no such thing as Descartes' *"res extensa"*.

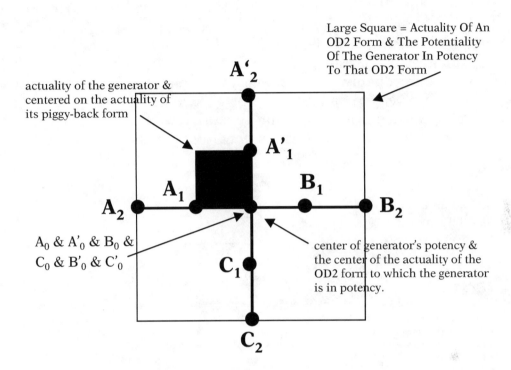

DRAWING #13: GENERATOR WHOSE ACTUALITY = .5A + .5A' & IN POTENCY TO A FORM CURRENTLY AT OD2. ONTOLOGICAL QUANTUM = ½

E: THE DIFFERENCE BETWEEN FORMS & GENERATORS:

Let's now concentrate on spelling out: (1) the difference between forms and generators; (2) how every generator is concentric with one piggyback form; and (3) how space is logically extended.

Drawings #13 & 14 represent the fact that the center of the *generator's* actuality invariably moves away from the center of its potentiality, since generators never *simultaneously* actualize more than 3 of their 6 potencies—no 2 of which are ever mutual opposites. Drawing #12 represents the fact that the center of every *form's* actuality invariably stays put at the center of its potentiality, since every form always actualizes all six of its potencies equally. Because of this absolutely essential difference (and to nail it in place in our minds), we can refer to the generators as *tri-active* hebdomads and refer to the forms as *hex-active* hebdomads.

Because of that absolutely essential difference, no piggyback form can, by itself alone, ever go anywhere away from the center of the universe. How, then, do piggyback forms move around in the universe?

Viewing Drawings 12-14, isn't the answer to that rather obvious? Take #12, and affix the center of its black box to the center of either one of the black boxes in #13 & #14. Once you say the center of that form's actuality and the center of that generator's actuality must forever remain logically concentric, you now have a piggyback form able to go anywhere in the universe its carrying generator can go.[1]

Also, you've now got two very different kinds of *basic realities* joining to produce an *ultimate constituent* of the universe. This duality, more so than its two hebdomads, is an ultimate constituent for 2 reasons: (1) No matter how much logic allows you to break the duality apart *mentally*, you cannot break it apart *de facto*; and (2) no generator is ever separated from a piggyback form, and no form (except for angels) is ever separated from a carrying generator; and so, no ultimate *constituent* is ever merely a basic *reality* (*i.e.:* form or generator) by itself alone.

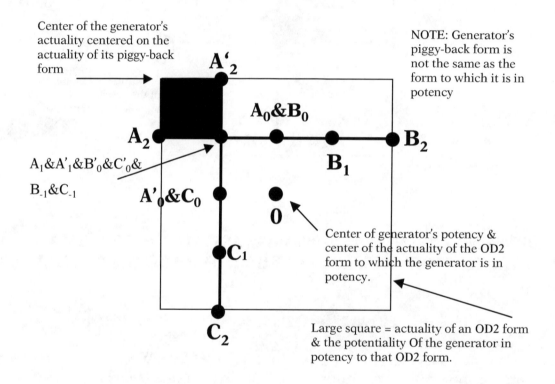

Center of the generator's actuality centered on the actuality of its piggy-back form

NOTE: Generator's piggy-back form is not the same as the form to which it is in potency

A'_2

$A_0 \& B_0$

A_2

B_2

$A_1 \& A'_1 \& B'_0 \& C'_0 \& B_{-1} \& C_{-1}$

B_1

$A'_0 \& C_0$

0

C_1

Center of generator's potency & center of the actuality of the OD2 form to which the generator is in potency.

C_2

Large square = actuality of an OD2 form & the potentiality Of the generator in potency to that OD2 form.

DRAWING #14: GENERATOR WHOSE ACTUALITY = (1A -.5B) + (1A'-.5C) & IN POTENCY TO AN OD2 FORM. ONTOLOGICAL QUANTUM = ½.

Finally, you've now got two kinds of *hebdomads* combining to produce a *quartuordecimad*. As you possibly know, "*quartus-decimus*" is the way one says 14 in Latin. And who cannot readily see that, when you add together 2 basic realities each consisting of 7 factors, the result is an ultimate constituent consisting of 14?

---

[1] In Esoptrics, there are forms whose actuality remains centered on the center of the universe due to the fact that they are not logically concentric to a generator. They're called "angels". Contrary to what Science imagines, there is no such thing as a "theory of everything", if it does not also account for how angels fit into our universe. On that point, Esoptrics is way ahead of Science.

If all of that is reasonably clear, we can now turn to the task of using Geometry to allegorize graphically how space is logically extended. To start, look at Drawings #13 & #14 and ask: "What 6 potencies is the generator using in the course of its travels? Is it, too—as with its piggyback form—in potency to infinity?" It most certainly is not. Every generator always and everywhere remains in potency to the actuality of some form which is other than its own piggyback form. That has to be because, if the clear areas in #13 & #14 are the clear areas in #12, then the center of the generator's actuality is not remaining glued to the center of the actuality in #12. It has to be, then, that the center of any given generator's *potentiality* is the center of the actuality of a form other than that given generator's piggyback form. What form *is* that other form?

A'

1	2	3	4	5	6	7	8
9	10	11	12	13	14	15	16
17	18	19	20	21	22	23	24
25	26	27	28	29	30	31	32
33	34	35	36	37	38	39	40
41	42	43	44	45	46	47	48
49	50	51	52	53	54	55	56
57	58	59	60	61	62	63	64

A   B

C

DRAWING #15: THE ACTUALITY OF AN OD4 FORM &, THEREFORE, THE POTENTIALITY OF ANY GENERATOR IN POTENCY TO IT. ONTOLOGICAL QUANTUM = ¼.

For any given generator, the current ontological distance of its piggyback form determines what must be the current ontological distance of the form to which that generator shall be in potency. I will not, however, go into that beyond saying this: Increase the current OD of a given piggyback form, and its carrying generator must become in potency to a form whose current OD is higher than that of the form to which the generator was previously in potency. As a result, you will increase the range and speed of the generator's locomotion. To say more than that would make this letter much too long, and I don't see any need for such detail, since the meaning of logically extended space can be explained without such detail. Drawing #15 (prior page) shall now help us do that. Here again, of course, I'm dealing with the need to work with a square when I really need a cube.

F: LOGICALLY EXTENDED SPACE:

Let's refer to the whole of Drawing #15 as a matrix. We can then say that each of the 64 squares in the matrix represents a possible primary act which can be described in the terms of which quantity of which potency is being either activated or transferred. I'm not about to describe all 64 of the squares, but will describe several. As we describe the squares, the meaning of "transferred" shall become clear.

In case you've not figured them out for yourself, let me pause to give you a detailed description of each of the 6 potencies available to every generator. We'll take them 2 mutually opposed potencies at a time.

What I call the "A potency" is the left half of the matrix, including the left half of the *cube* we must *imagine*. To be more technically exact about it, the A potency is everything on the A side of the axis A'C. Oppositely, what I call the "B potency" is the right half of the matrix, including the right half of the cube we must imagine. To be more technically exact about it, the B potency is everything on the B side of the axis A'C. To be even more technically exact about these 2 potencies, the A potency is everything on the A side of the plane formed by the axes A'C & B'C', and the B potency is everything on the B side of that plane. That plane can be imagined as a flat disc perpendicular to the AB axis at its center.

What I call the "A' potency" is the rearward half (*i.e.:* the half running from the horizontal center of the matrix toward the top edge of the page), including the rearward half of the cube. To be more precise about it, the A' potency is everything on the A' side of the axis AB. Oppositely, what I call the "C potency" is the forward half (*i.e.:* the half running from the horizontal center of the matrix toward the bottom edge of the page), including the forward half of the cube. To be more exact, the C potency is everything on the C side of the axis AB. To be even more technically exact about these 2 potencies, the A' potency is everything on the A' side of the plane formed by the axes AB & B'C', and the C potency is everything on the C side of that plane. That plane can be imagined as a flat disc perpendicular to the A'C axis at its center.

As for what I cannot depict, the B' potency is the half of the cube we must imagine as rising from the surface of the page, and the C' potency is the half of the cube we must imagine as protruding downward from the backside of the page. To be more exact, the B' potency is everything on the B' side of the plane formed by the axes AB & A'C, and the C' potency is everything on the C' side of that plane. That plane can be imagined as a flat disc perpendicular to the B'C' axis at its center.

As you can see, some of the potencies overlap. Such, however, is not true of the mutual opposites A & B or of the mutual opposites A' & C or of the mutual opposites B' & C'. .

In #15, pg. 107, square 28 = .25A + .25A'. If square 28 is imagined as a cube on the *first* tier *above* the mid plane of the matrix, then square/cube[1] 28 = .25A + .25A' + .25B'. If it's on the *first* tier *below* the mid plane, 28 = .25A + .25A' + .25C'.

Performing any one of those primary acts, the generator fills the appropriate box with a unit of force (as in "tension" or "vibration") which is matter in the strictest, truest, primary, most definite, and unqualified sense of the word "matter". For, when we experience our sense images (and these we deem to be, at least in some cases, intra-mental representations of the internal properties of the matter comprising the extra-mental, material objects affecting us), what our senses are doing is this: feeling the units of force (*i.e.:* friction?) the generators generate. Force is the "stuff" of *sensible* matter but not of anything else. Since it must be generated, it's a kind of "stuff" *parasitic* unto its generator and is created when generated but destroyed when not generated and then replaced by a newly generated unit.

Square 27 (#16, next page) = (.5A -.25B) + .25A'. The "-.25B" (Notice the minus sign) signifies that .25A has been transferred over to the B potency. How so?! Remember the technically exact description we gave to the B potency: It's everything on the B side of the axis A'C. Now look again at Drawing #14 (pg. 106), and notice how the axis A'C has shifted to the left and, thereby, has increased the extensiveness of the B potency and diminished that of the A potency. It's that simple.

If square 27 is imagined as being a cube on the *second* tier *above* the mid plane of the matrix, then cube/square 27 = (.50A -.25B) + .25A' + (.50B' -.25C'). The "-.25C'" signifies that .25B' has been transferred over to the C' potency. To represent that graphically, I would have to draw a cube in which the plane formed by the axes AB & A'C has shifted upward along the axis B'C' and has, thereby, increased the extensiveness of what's below that AB/A'C plane.

Performing either one of those primary acts, the generator fills the appropriate box with a unit of force which is matter in the strictest, truest, primary, most definite, and unqualified sense of the word "matter". For, when we experience our sense images, we sense units of force generated by generators.

Square 26 (Drawing #17, pg. 111) = (.75A -.5B) + .25A'. The "-.5B" signifies that .5A has been transferred over to the B potency. Drawing #17 illustrates that by showing squares 27 & 28 now on the B side of the axis A'C. Performing this act, the generator fills the appropriate box with a unit of force/vibration. Pardon me, if I decline to repeat this verbal bit about units of force in the following descriptions.

Square 25 (Drawing #18, pg. 112) = (1A -.75B) + .25A'. If square 25 is seen as a cube in the *first* tier *below* the mid plane of the matrix, then cube/square 25 = (1A -.75B) + .25A' + .25C'.

Square 17 (Drawing #19, pg.113) = (1A -.75B) + (.50A' -.25C). If square 17 is seen as a cube in the *first* tier *above* the mid plane of the matrix, then cube/square 17 = (1A -.75B) + (.50A' -.25C) + .25C'.

Square 64 (Drawing #20, pg. 114) = (1B -.75A) + (1C -.75A'). If it's seen as a cube in the top most tier of the matrix, then 64 = (1B -.75A) + (1C -.75A') + (1B' -.75C').

---

[1] Remember: It's a square in the drawing, but a cube in the imagination's more accurate rendition of how a geometrical allegorization of it should represent the generator's act.

A'

$A_1 \& B_{-1} \& A'_0 \& C_0 \& B'_0 \& C'_0$

1	2	3	4	5	6	7	8
9	10	11	12	13	14	15	16
17	18	19	20	21	22	23	24
25	26		28	29	30	31	32
33	34	35	36	37	38	39	40
41	42	43	44	45	46	47	48
49	50	51	52	53	54	55	56
57	58	59	60	61	62	63	64

A   **B**

C

DRAWING #16: GENERATOR WHOSE ACTUALITY = (.5A -.25B) + .25A', & IN POTENCY TO AN OD4 FORM. ONTOLOGICAL QUANTUM = ¼.

If you notice, every change from one square to an *adjacent* one involves a change of exactly 25% in some *one* of the generator's potencies. That 25% is what we can call the "ontological quantum" binding on all generators traveling within the confines of the actuality of a form currently at OD4. Always and everywhere, the ontological quantum equals the inverse of the current OD of the form to which a generator is in potency. For example, if the generator is in potency to a form currently at OD4, the ontological quantum is ¼; if the form is currently at OD5, the ontological quantum is 1/5; at OD6 = 1/6; at $OD2^{128} = ½^{128}$ etc.. Due to that rule, every primary act on the part of any and every generator (and every unit of force produced by every such act) is always roughly 10^{-47} cm., if we describe every such act in the terms of Science's "Geometry speak".

As long as someone or something makes it an inviolable law that every change from primary act to primary act must adhere to the pertinent ontological quantum, then the matrix shows us which acts are adjacent to one another by virtue of the logical necessity imposed by that inviolable law of the ontological quantum. Of course, if that law of the ontological quantum is relaxed, then here's an example of what that would entail: A generator performing the act at square #64 can change from that act directly to the act characteristic of square #1, and that would instantly put it at square #1. *Relax* the law of the ontological quantum, and every square in the matrix is logically adjacent to all the other squares.

$A_2 \& B_{-2} \& A'_0 \& C_0 \& B'_0 \& C'_0$

**DRAWING #17: GENERATOR WHOSE ACTUALITY = (.75A -.5B) + .25A',
& IN POTENCY TO AN OD4 FORM. ONTOLOGICAL QUANTUM = ¼.**

Nevertheless, *enforce* the law of the ontological quantum, and here's an example of what they would entail: Every generator performing the act characteristic of square #8 cannot change from that act to any other save either the act characteristic of square #7 or the act characteristic of square #16. I leave out square #15 because, to this day, I *suspect* the law of the ontological quantum rules out diagonals by limiting generators to X% of *one* potency per change from primary act to primary act. A move from 8 to 15 takes a 25% change at once in both A' and A.

A'

$A_3 \& B_{-3} \& A'_0 \& C_0 \& B'_0 \& C'_0$

1	2	3	4	5	6	7	8
9	10	11	12	13	14	15	16
17	18	19	20	21	22	23	24
■	26	27	28	29	30	31	32
33	34	35	36	37	38	39	40
41	42	43	44	45	46	47	48
49	50	51	52	53	54	55	56
57	58	59	60	61	62	63	64

A   B

C

**DRAWING #18: GENERATOR WHOSE ACTUALITY = (1A -.75B) + .25A',
& IN POTENCY TO AN OD4 FORM. ONTOLOGICAL QUANTUM = ¼.**

For Leibniz, when an ultimate goes from one box[1] in the above matrix to another, it means God has ceased to create that ultimate in the first box and has commenced to create it in the next. It's called "transcreation". That is by no means true for Esoptrics. For Esoptrics, when an ultimate goes from one box to another, it merely ceases to perform one kind of primary act and commences to perform one which—from the standpoint of what's logically necessary according to the law of

---

[1] Leibniz, of course, would not say "box"; he'd say "place in space".

the ontological quantum—is logically adjacent to the prior one. Here again, then, to speak of Leibniz and Esoptrics in the same breath is ludicrous.

A' $A_3\&B_{-3}\&A'_1\&C_{-1}\&B'_0\&C'_0$

1	2	3	4	5	6	7	8
9	10	11	12	13	14	15	16
	18	19	20	21	22	23	24
25	26	27	28	29	30	31	32
33	34	35	36	37	38	39	40
41	42	43	44	45	46	47	48
49	50	51	52	53	54	55	56
57	58	59	60	61	62	63	64

A   B

C

DRAWING #19: GENERATOR WHOSE ACTUALITY = (1A -.75B)+(.5A'-.25C), & IN POTENCY TO AN OD4 FORM. ONTOLOGICAL QUANTUM = ¼.

Of course, Esoptrics would add this: No generator can move *itself* to change from primary act to primary act. It must be moved to do so by its piggyback form or by collision with another generator. Even in that latter case, though, the collision occurs only because of what each colliding generator's piggyback form was moving it to do; and so, in the final analysis, no generator changes from primary act to primary act save to the extent a piggyback form—either directly or thru another generator—moves it to change. But, then, no form can change from primary act to primary act unless moved to do so by the extra-cosmic influence. That, though, is hardly Leibniz; rather, it is a principle which was dear to Aristotle and St. Thomas Aquinas. It's the ancient principle which says that there can be no movement save where The Prime Mover, God, causes the movement either immediately or thru an intermediary.

1	2	3	4	5	6	7	8
9	10	11	12	13	14	15	16
17	18	19	20	21	22	23	24
25	26	27	28	29	30	31	32
33	34	35	36	37	38	39	40
41	42	43	44	45	46	47	48
49	50	51	52	53	54	55	56
57	58	59	60	61	62	63	

$A_{-3} \& B_3 \& A'_{-3} \& C_3 \& B'_0 \& C'_0$

DRAWING #20: GENERATOR WHOSE ACTUALITY = (1B -.75A)+(1C -.75A'), & IN POTENCY TO AN OD4 FORM. ONTOLOGICAL QUANTUM = ¼.

In Drawing #15, the point of perhaps crowning importance is this: Each primary act is a perfectly unique combination of potency kind(s) and quantity(ies). As long as that unique combination remains as it is, it is an act which remains the same and which undergoes not the slightest trace of any kind of internal change whatsoever. As long as that is so, that unique combination is an act which is not involved in *time*; instead, it is an act involved in *changeless duration*.

That does not mean the act is not involved in *history*, which is to say involved in the universe's progress from the start to the finish of what Esoptrics calls the universe's "ninth epoch". As history moves thru the ninth epoch, it carries each generator and form with it just as *steadily*[1] as it does any other generator and form. It's just that, as history does so, it, so to speak, switches some from one hand to another more frequently than it does others; and so, for one form, the entire journey takes only an astronomically tiny fraction of a second; for a vast multitude of others, it takes trillions of trillions of eons; and, for us, it takes 18 trillion years.

Now, here's an interesting development. In the midst of trying to explain Drawing #15 to you, I suddenly achieved an important advancement in Esoptrics. I have often wondered: Must every form's every primary act involve the activation of all 3 sets of 2 mutually opposed potencies, or can they sometimes activate only 1 or 2 sets of 2 mutually opposed potencies? I wondered about that because a hunch always insisted generators must be able to generate either 1 or 2 or 3 dimensional units of force, but could never figure out how that could be possible, unless the *forms* sometimes activated either 1, 2, or 3 sets of 2 mutually opposite potencies. Writing this for you, I suddenly realized how the generators could do such—namely: Generators are not limited to actualizing only the *cubes* in the matrix. It was merely a case of being too fixated on the *cubes* to notice the lines and the squares staring me in the face. Though I saw it not when I first drew them, Drawings #13 & #14, graphically represent how a generator can actualize squares rather than a cube, and Drawing #21 on the next page, graphically represents the actualization of a line.

As for one-dimensional acts on the part of the generator in Drawing #15, it's possible to have primary acts such as: (1) .25A, (2) .5A -.25B, (3) .75A -.5B, (4) 1A -.75B, (5) .25B, (6) .5B -.25A, (7) .75B -.5A, (8) 1B -.75B, (9) .25C, (10) .5C -.25A', etc., and each of these acts would produce a one-dimensional act and a one-dimensional unit of force our senses would detect as a line. As for two-dimensional acts on the part of the generator in Drawing #15, it's possible to have acts such as: (1) .25B' + .25A, (2) [.5B' -.25C'] + .25A, etc., and each of these acts would produce a two-dimensional act and a two-dimensional unit of force our senses would detect as what we might call a "chip" or a "sheet". That being the case, I am now quite decided that every act of every form always involves the activation of all 3 of the 3 sets of 2 mutually opposed potencies.

I think that's enough said to make it adequately clear what logical extendedness is. Seeing what logical extendedness is, it should also be adequately clear that—while it definitely says *physically* extended space is not real—Esoptrics is far from saying that space is not real. The issue is not whether or not space is real; rather, the issue is *what kind* of space is real. For Esoptrics, every generator is that which navigates thru the only real kind of space, because, in its actuality, every piggyback form is that which provides the generators with a truly real kind of *logically* extended space thru which they can steer their courses. And this, according to you, dear professor, is merely the same as what was said by "Leibniz and other members of the 7th c. rationalists school"???!!!

It now remains for me to go into more detail regarding the origins of the 6

---

[1] I say "steadily" rather than "continuously" because Esoptrics says that, below roughly 10^{-95} seconds, nothing anywhere in the universe undergoes any kind of change whatsoever; and so, below roughly 10^{-95} seconds, the universe makes no progress thru the ninth epoch.

potencies. On what grounds does Esoptrics justify its appeal to an extra-cosmic influence? On what grounds does Esoptrics justify its assertion that this extra-cosmic influence is sexpartite? How it would delight me to launch into those explanations here and now! I must, however, resist the inclination to do so, because, at 18 pages in length, this letter is already uncharitably long. Then too, I do not even know for sure that you are receiving my letters. And, if you are receiving them, are you reading them? And, if you are reading them, are you getting anything out of them other than a reason to reject everything I say? How encouraging it would be were you to send me at least a brief note assuring me you are indeed receiving, reading, and profiting from, my letters!

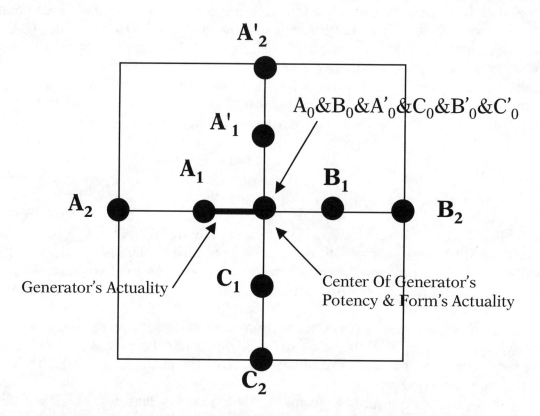

The diagram shows points labeled: A'_2 (top), A'_1, A_1, B_1, A_2, B_2, C_1, C_2, with the center labeled $A_0 \& B_0 \& A'_0 \& C_0 \& B'_0 \& C'_0$. Labels: "Generator's Actuality" and "Center Of Generator's Potency & Form's Actuality".

DRAWING #21: A GENERATOR WHOSE ACTUALITY = .5A & IN POTENCY TO AN OD2 FORM. ONTOLOGICAL QUANTUM = ½

It is, however, by no means *necessary* for that to happen. Whether it happens or not, I shall, in a third letter, spell out the introspective ground work which led to the logic of the mirror, and its conclusion that there is a sexpartite extra-cosmic influence which imparts to the forms a logical extendedness *and various patterns of rotation*. I shall do so because, after—for your privacy's sake—stripping from my letters to you all reference to your last name and the university at which you teach, I intend to mail copies of them to every one of the 1,724 departments of Philosophy and Physics on my mailing list. In short, these letters are not for you alone.

Therefore, most generous and patient Sir, unless you order me not to, I shall finish and send you a copy of the intended third letter.[1] Until then:

May the Lord bless you and keep you,
And lift up His Countenance upon you.
Yea, may He make His Face to shine upon you,
And give you Peace.

In Christ Jesus,
    Thru Mary
        With Joseph,
A Brother To All:

EDWARD N. HAAS

שָׁלוֹם

Completion Date: Wednesday, August 29, 2007.
In Die Festo In Martyrio Johannis Baptistae, Consummatum Est.

---

[1] **NOTE OF JULY 12, 2009:** That letter was written but never sent to him. It follows as an essay rather than as the letter it was originally intended to be.

In physics, the smallest doll [He means smallest ultimate constituent of the universe – ENH] is called the Planck length. To probe to shorter distances would require particles of such high energy that they would be inside black holes. We don't know exactly what the fundamental Planck length is in M-theory, but it might be as small as a millimeter divided by a hundred thousand billion billion billion [1 followed by 32 zeros – ENH]. We are not about to build particle accelerators that can probe to distances that small. They would have to be larger than the solar system, and they are not likely to be approved in the present financial climate.

——**STEPHEN HAWKING:** *The Universe In A Nutshell*, pg. 178 as published in hardbound copy by Bantam Books, New York, November 2001. ©2001 by Stephen Hawking. [In other words, there's no way Science can observe whether or not Esoptrics is correct when it says the universe's ultimate microscopic constituents are 10^{-46}mm. in LB&D. After all, it takes us another 14 zeroes beyond the 32 already requiring a particle accelerator larger than the solar system.]

Document #6:

ESOPTRICS' BASIS IN PURELY INTROSPECTIVE REASON:

PART I: A DESCRIPTIVE LIST OF WHAT TARGETS I OBSERVE WITHIN THE CONFINES OF MY OWN ACTS OF CONSCIOUSNESS:

A: THE DIRECT ENCOUNTEREDS:

Whenever I am awake, I am, on one hand, aware of an immense series of sense images and, on the other, aware of myself as that which is aware of those sense images. I use the plural in referring to sensation because there is more than one *kind* of sensation. For example, my most noticeable sense images are colored shapes; but, others are sounds or odors or flavors or tactile sensations such as cold, heat, hardness, softness, roughness, joy, sorrow, equilibrium, and so forth.

No matter what I may do, I cannot—as long as I am awake—escape this duality of: (1) an observer who feels and (2) the sense images he feels. In effect, the feeling observer is one side of a coin and his sense images the opposite side. I say that to emphasize a point: As with the two sides of the same coin, I, the feeling observer, and my observed sense images cannot be disjoined, which is to say I cannot separate them in *fact* no matter how extensively I may do so in *thought*.

The feeling observer, I label: "the encounterer"; the sense images *currently* being felt by the observer, I name: "encounter*eds*". Sense images *not currently* but *able* to be encountered, I call: "encounter*ables*". The encounterer's act of feeling the encountereds, I dub: "encountering". Having done that, I can then say this: For myself, the most obvious and persistent of all observable facts is: An encounterer is encountering one or more encountereds.

Of my sense images, I have a very direct and concrete awareness. To express that fact, I refer to my sense images as "*direct* encountereds".

At least in the vast majority of my *visual* images, the direct encountereds present me with extremities—whether in space or in time or in both. Regardless of whether or not or why it's legitimate to do so, I can use those spatio-temporal extremities as grounds for distinguishing one direct encountered from another. I am able to do that, of course, because: **FIRST:** I have a power of attention which I can willfully cause to *contour* the various sets of extremities; **SECOND:** I am then able to notice the difference between: (1) the way I must will to cause my power of attention to *contour* one given set of extremities, versus (2) the way I must will to cause my power of attention to *contour* some other given set of extremities.

Having done that, I notice this: Within its extremities, each of the direct encountereds presents me with its internal characteristics. I refer to these internal characteristics as the "encounterableness of the direct encountered". If speaking of the internal characteristics of sense images *not currently* being—but *able* to be—encountered, I refer to the "encounterableness of the direct encounter*ables*".

Despite what I just said, I shall often use the term "direct encounterables" to refer to any and every sense image whether currently being encountered or not. Pardon me for that, I beg you. Such a use shall save us much time and space. Hopefully, you'll be able to tell from the context which way the term is being used at the moment.

I divide the direct encounterables into two kinds: (1) those whose encounterableness is *perfectly* subject to my whims, and (2) those whose encounterableness is either *im*perfectly subject to my whims or—as far as I can tell—not subject at all. The former, I call by such names as "intra-mentally controlled", "pliant", "imaginary", "whimsical" or "subjective" direct encounterables; the latter, I call by such names as "extra-mentally controlled"[1], "intrusive", "defiant" or "objective" direct encounterables.

When *imaginary*, the encounterableness of a direct encountered is *perfectly voluntary*. When *defiant*, the encounterableness of a direct encountered is either *im*perfectly voluntary (*i.e.: im*perfectly *in*voluntary) or *perfectly in*voluntary. For example: (1) As far as I can tell, I have *no* control *whatsoever* over what I shall see when I unobstructedly view the Sun, because I have no way to modify whatever forces me to see what I see when looking at the Sun unobstructedly; (2) I have *some* control over what I shall see when I unobstructedly view this piece of paper, because, though I must move my hands to do so, I can crumple this sheet and, thereby, modify whatever forces me to see what I see when looking at it unobstructedly; but, (3) I have *total* control over what I shall see when I unobstructedly view what I shall write on this sheet of paper, because nothing but my own will determines what I shall see when looking at what to write.

B: THE REFLEX ENCOUNTERED

So then, in encountering the involuntary direct encountereds, I come face to face with an immense series of internal characteristics which are perfectly or imperfectly defiant. Such could hardly be said of that side of me which is performing this marvelous act called "encountering the direct encountereds". I encounter not a one of the encounterer's internal characteristics. Not only do I *not* encounter the encounterer; but, rather, I can't even determine for sure whether or not it *has* an encounterableness. In other words, I don't know for sure whether or not it is possible for me, the encounterer, to come face to face with the internal characteristics of my encoun-

---

[1] They are extra-mentally controlled in the sense that, no matter how hard I look, I can find nothing within my mind itself able to control them. In such a case, of course, "mind" means my self in so far as I am *aware* of my self. That being the case, extra-mentally controlled direct encounterables are still *extra*-mentally controlled no matter how extensively they might be controlled by some part of myself of which I am unaware.

tering self. Naturally, if there are not, and never will be, any circumstances under which I, the encounterer, can come face to face with my internality, then my internal characteristics can never *rightly* be referred to as an encounter*able*ness. I strongly *suspect*, of course, that there <u>are</u> circumstances under which I, the encounterer, can turn my own self into a *direct* encountered; however, I have not the foggiest notion regarding what those circumstances might be, save for this: There would need to be The Infinite One Who, by boosting me to infinity, causes me to perform an act of encountering at infinite velocity. More on that later!

What, then, shall we say as long as there is no encounter of the encounterer?! Is the encounterer *nothing* to itself? Have I no awareness of my self whatsoever? By no means! If I am not *directly* aware of my self, I am, nevertheless, <u>in</u>directly aware of the *fact* of my presence. Such indirect awareness of my self is the result of being aware there is *something* feeling sense images and moving its power of attention around among them. Indeed, so keenly am I aware of what that something is doing to the sense images, I am pressed to the wall with longing to whirl around and to glimpse the internal characteristics of whatever is feeling the sense images. To put it another way, the encounterer—*without encountering <u>himself</u>!*—encounters his *act* of *encountering* the direct encountereds. So intently does he encounter that act, he is obsessed with the idea of encountering whatever is performing that act.

What does that do to my act of encountering the direct encountereds? Does it transform that act itself into a *direct* encountered? Not at all! The act of encountering the direct encountereds is encountered, so to speak, <u>on the rebound</u>. For that reason, it seems best to refer to the act of encountering the direct encountereds as a *"reflex"* encountered. Say that, and I can now say the second of my most obviously and persistently observable facts is: The encounterer is directly encountering the direct encountereds and reflexively encountering the reflex encountered.

Wait a minute! When the encounterer reflexively encounters the reflex encountered, is he aware he is doing so? If so, what shall we say in that case? Shall we say the encounterer is—in a doubly reflexive manner—encountering a doubly reflex encountered?! I suggest we shall say no more than what we've already said—namely: The encounterer is reflexively encountering the reflex encountered.

C: THE TECHNOLOGICAL UNENCOUNTERABLES

We now have an encounterer directly encountering whimsical and defiant direct encountereds and reflexively encountering reflex encountereds. The encounterer himself is neither directly nor reflexively encountered. He is the <u>un</u>encountered.

He is also the unencounter<u>able</u> in the sense that we do not currently know how to convert the encounterer into a direct encountered. We cannot refer to the encounterer as *absolutely* unencounterable for a very simple reason: We do not know for sure that there are <u>no</u> circumstances under which one can come face to face with the encounterer's encounterableness.

Possibly, we can best express that fact by referring to the encounterer as a "technological" unencounterable. Using the adjective "technological" implies that—

under the <u>current</u> level of our <u>technology</u>!—the encounterer is unencounterable. At the same time, though, it leaves open the question of whether or not someone will *eventually* introduce a level of technology able to render the encounterer a direct encountered.

The instant we refer to the encounterer as a technological unencounterable, we vigorously suggest this question: Is the encounterer the *only* technological unencounterable? One does not have to look far to find an evidence suggesting a negative answer. How is that?

Who or what forces upon the encounterer the defiant encounterableness of the perfectly and imperfectly involuntary direct encountereds? Certain it is that I do not begin to know all I desperately long to know concerning what determines the internal characteristics of my involuntary direct encountereds. Where should I look to find that knowledge?

Were we to come face to face with all the internal characteristics of the encounterer, would each of us find that his own self is the *sole* cause of the encounterableness of every one of his encountereds? I, for one, know not how one could answer with *certitude*[1] one way or the other. Be that as it may, I am absolutely certain I myself encounter far more evidence for a negative response than I do for an affirmative one. Therefore, I am absolutely certain it is at least a well founded opinion to say that the encounterer is <u>not</u> the <u>only</u> technological unencounterable influencing the defiant encounterableness of my involuntary direct encountereds.

D: THE CONTACTABLES

How shall we distinguish between the encounterer and all other technological unencounterables? Let's say that, with regard to himself, every encounterer is the *reflex* technological unencounterable, and all other technological unencounterables are—with regard to the encounterer—*direct* technological unencounterables.

The direct technological <u>un</u>encounterables can also be referred to as the "contactables". They can be labeled that way for a simple reason: It is at least a well founded opinion to say that, though the direct technological unencounterables "touch" their extremities to the extremities of the encounterer's act of encountering, nothing of what is inside of their extremities enters to any extent into what is en-

---

[1] Certitude = a fearless conviction whose fearlessness is either absolutely or practically justified by evidence. The fearlessness is *absolutely* justified if my conviction pertains to what is empirical, which is to say what I am certain of is an observable fact which is "theory neutral", which is to say in no way influenced by entrenched prejudices, suppositions, inferences, and the like. The fearlessness is *practically* justified if it pertains to an inference corroborated by a massive preponderance of empirical and/or inferential evidences. If the preponderance is less than massive, my conviction is fearful and, as such, is merely a well-founded opinion. I claim no certitude with regard to whether or not my *words* accurately and fully describe exactly what I seek to describe. I especially claim no certitude with regard to whether or not my words shall convey to others even so little as *most*, let alone *all*, of what I seek to convey. Over the decades, I have received a few replies to my writings. The vast majority of them ever left me dumbfounded over how grossly their authors misunderstood and distorted my words. I, it seems, have little or no capacity to choose words able to convey my thoughts. Hence my lack of even so little as a well-founded opinion regarding what my words shall say to others.

closed within the extremities of the encounterer's act of encountering.

Did that go over your head? Well, to re-express it in a simpler manner, imagine two steel balls which very often bump up against each other but without ever penetrating one another to any extent whatsoever. For such steel balls, each is a contactable with regard to the other.

There's an important reason for drawing out that point. As we shall soon see, it is important to realize that no contactable is to any extent, or in any sense, actually present *inside* the encounterer's act of encountering; the encounterer, though, <u>is</u>. The encounterer is not present inside the act of encountering **_as an encountered_**; nevertheless, he is most certainly present inside that act **_as its agent_**.

Did that, too, go over your head? Here again, then, let's re-express it in a simpler way. Imagine, then, that the encounterer is something living at the center of one of the two steel balls just mentioned. Each time the two steel balls bump one another, the encounterer feels the "shock waves" generated inside of his own steel ball as a result of the collision, but never either *feels* himself or realizes any other method of experiencing his own encounterableness. At the same time, though, neither does he ever feel, or in any other way experience, the other steel ball itself. Indeed, he doesn't even feel, or in any other way experience, the internal characteristics of the *steel* inside of the extremities of his own sphere. As far as forms of awareness are concerned, all he ever does is: (1) feel (**_i.e.:_** directly encounter) the "shock waves" generated inside of his own steel ball as a result of colliding with the other steel ball, and (2) experience in a <u>non-feeling</u> way (**_i.e.:_** reflexively encounter) his act of feeling those "shock waves".

E: THE ARBITRIA

Let's now take a closer look at the encounterer. We must do so because the encounterer does far more than merely encounter the direct encountereds. He also focuses his power of attention upon them. What does that mean?

Hold this entire page in your field of vision. Notice you can fix your eyes on one word and, yet, keep the entire page in sight. It's clear, then, you have a "power of *attention*" quite distinct from your power of *vision*. In fact, the power of *attention* can move around within any framework presented by the power of *vision*, and it can do so without making the slightest change in that framework's extremities.

Continuing to hold the entire page in your field of vision, move your power of attention around within that framework, and locate at least two instances of the letter "s." Now that you've done that, *how* did you do it?

I say this: By means of several acts of will, you both moved your power of attention around within the framework of visual sensations and *contoured* your attention to the letters found there. From past experience, you already knew: (1) what choices would cause your power of attention to scan a page, and (2) what choices would *contour* the power of attention to the sense image signified by the phrase "the letter 's'." Therefore, wherever you found contouring some letter required the acts of will you previously associated with the letter "s," you confidently concluded the target of your attention was an instance of the letter "s".

For example's sake, imagine how a blind sculptor works. First, he wills his hands to feel his subject's exact shape. Next, he moves his hands to clay and exactly recreates that unique shape. It's a procedure clearly bespeaking, in the sculptor's mind, a concept comprised of his awareness of the choices associated with making his hands *contour* the exact shape peculiar to his model. To me, that suggests this: Unlike animals with a consciousness unable to *reflect back* upon itself, each of us is able to *think* about his sensations and to *express* them in the terms of:

THE ACTS OF WILL HE MUST PERFORM
IN ORDER TO CONTOUR HIS POWER OF ATTENTION
TO THEIR EXTREMITIES.

The encounterer, therefore, does far more to the direct encountereds than encounter them: He *expresses* them in the terms of his own will's activity. That means this: By means of contouring his power of attention, he attaches to the target of that contouring operation a particular kind of reflex encountered which we can call "actual thoughts" or "actual concepts". Each of these actual thoughts (*i.e.:* actual concepts) is that series of "attention focusing acts of will" which brought about the contouring and/or *locating* of some particular sense image. To save space, we can replace the phrase "attention focusing acts of will" with the term:

THE ARBITRIA.

Therefore, the encounterer expresses his encountereds in the terms of his arbitria. He does so by somehow forming a permanent link between any given encountered and the arbitria he uses to contour and/or locate it. Doing that, he delineates that encountered either in spatio-temporal terms or in the purely logical terms of a chart of classifications.[1]

---

[1] "Intentionality" started with the Scholastics (*i.e.:* Catholic Philosopher-Theologians of perhaps the 9th to the 17th century who, somewhat like "Bible-only-devotees", advocated intellectual progress thru dialectics, which is to say thru debates over inferences drawn from verbal gymnastics rather than from, as Aristotle put it, "keenly focusing upon what the external world presents to us"); but, even today, there's much talk about "intentionality". It has to do with "mental states" (I would say encountereds.) which exhibit "aboutness". It's a way of saying that, by their very nature, at least some of the targets we observe in our heads automatically direct our attention to something else. That, of course, is exactly what the arbitria do. From that, one might think that, in writing about the arbitria, I'm merely repeating what others have said about intentionality. Think that, and you shall be sadly mistaken. Nowhere, that I know of, do those who write about intentionality ever mention the acts of will whereby we contour our attention. They especially do not mention the notion that one can identify, know, and define this-or-that target of consciousness in the terms of the acts of will whereby we contour our attention to that target. That's a very sad state of affairs, because the arbitria are **the** means by which every intentional mental state *has* intentionality. They alone are what the encounterer uses to connect one encountered to another encountered or contactable. The arbitria are thus the sole and principle cause of all intentionality; and so, if any given individual says nothing of the acts of will whereby we contour our attention to this-or-that target of consciousness, that given individual knows nothing of intentionality's cause, and that clearly leaves his concept of intentionality with a giant hole in it. It's as if one went to a learned professor's lecture on mammals, and, at the end of the lecture one said to him: "In your list of mammals, you didn't mention elephants. Since they are my favorite mammal, I do wish you had mentioned them." To your dismay, he replies:

If you want an example of using the arbitria to delineate an encountered in spatio-temporal terms, then I would assure you of this: I know exactly how I must will in order: (1) to focus my power of attention upon a particular one of my book shelves, (2) to focus my power of attention upon a particular one of the places on that particular book shelf, and (3) to focus my power of attention upon that particular book entitled **Documents Of Vatican II**.

If you want an example of using the arbitria to delineate an encountered in the purely logical terms of a chart of classifications, then I would assure you of this: (1) In the arbitria which focus my power of attention upon the book shelf mentioned above, I know which arbitria enable me to classify that encountered as one of the many included in the classification "book shelf"; (2) in the arbitria which focus my power of attention upon a particular one of the places on that particular book shelf, I know which arbitria enable me to classify that encountered as one of the many included in the classification "spot on a book shelf"; and (3) in the arbitria which focus my power of attention upon a particular book occupying that particular spot on that particular shelf, I know which arbitria enable me to classify that encountered as one of the many included in the classification "book".

The phrase "Focus upon an actual thought/concept" implies a power of attention focused on the choice focusing it. To carry out such a *reflexive* task without the aid of sensation is at least exceedingly difficult, if not impossible for now, and that might explain why we cannot do even so little as *notice* our actual thoughts without the aid of sensation. Be that as it may, carry out that task, and what you shall attentively observe shall by no means be any kind of sensation.

F: OBJECTIVE VS. SUBJECTIVE ARBITRIA:

When I focus my power of attention upon an encountered, I can follow one of two courses: On the one hand, I can make a diligent effort to follow only those extremities which are defiant, which is to say those extremities which the direct encountered itself—if not some contactable by means of the direct encountered—*forces* upon me. On the other hand, I can make a diligent effort to ignore all such extremities in order to impose a set wholly and entirely the work of my own capriciousness.

Wherever I focus upon defiant extremities, the arbitria which focus the

---

"Elephants?! I've never heard of elephants. Is there such a mammal?" On the face of it, no matter how much he knows and says about *mammals*, that does not automatically give him specific knowledge of elephants. By the same token, examine the writings of those learned professors who expound at great length on intentionality, and you'll find they say nothing about the arbitria, and that's because no matter how much they know and say about intentionality, that does not automatically give them specific knowledge of the arbitria, and their failure to mention them is proof positive they have in fact never noticed the arbitria and know nothing significant about them. No matter how much the learned professor knows and says about mammals, his silence in their regard must eventually expose the fact he knows nothing specific about elephants, and no matter how much all the learned professors in the world know and say about intentionality, their silence in their regard must eventually expose the fact they've never noticed the arbitria and know nothing specific about them. Oh, and what a giant hole that bespeaks in their ability to observe their own minds!

power of attention are *objective*. After all, how I will focus my power of attention is—in this <u>*first*</u> case—at least primarily if not exclusively determined by what the encountered itself—if not some contactable by means of the direct encountered—presents to me. By way of example, the arbitria, which focus my power of attention upon the shape of my computer's monitor, are objective.

Wherever I focus upon capricious extremities, the arbitria, which focus the power of attention, are *subjective*. After all, how I will focus my power of attention is—in this <u>*second*</u> case—at least primarily if not exclusively determined by what extremities I myself arbitrarily choose to read into the encountered rather than by what that encountered itself presents to me. By way of example, the arbitria which focus my power of attention upon my computer as a machine out to kill me and take over my world, are subjective.

G: THE PRIME ABSTRACTABLE

Oh, the reflex encountered and its arbitria! What a marvelous experience! Do you begin to realize that? Are you not, then, keenly focused upon that reflex encountered and its arbitria? Are you so keenly focused upon it that it begins, so to speak, to move outward from the center, to merge with the direct encountereds and to become, as it were, a *direct* encountered always and everywhere enveloping all the rest of the direct encountereds?

Is there such a thing? Is there a single characteristic which—always and everywhere—envelops all the direct encountereds? There most certainly is a single characteristic which I can find in, and abstract from, every direct encountered at all times everywhere. It's one commonly called the "smoothness" and the "continuousness" of every direct encountered's flow thru time and/or space. Newton, I'm told, referred to it as a "fluxion". It is the most universal feature I find among my direct encountereds. If only in its method of passing through time, each and every direct encountered exhibits this characteristic. That is why I deem it quite proper to label it:

**THE PRIME ABSTRACTABLE =
THE CONTINUITY WITH WHICH
ALL DIRECT ENCOUNTEREDS
MOVE THRU TIME AND/OR SPACE.**

When first I noted the prime abstractable, it fascinated me no end. As I focused on it, all the problems of, and arguments against, the "continuum" raged fiercely in my mind, and I found myself pressed to conclude:

**NOTHING SIMILAR TO A DIRECT ENCOUNTERED
CAN ENGAGE IN THE KIND OF MOTION
OF WHICH CONTINUITY IS A CHARACTERISTIC.**

But, if no sense image moves *smoothly*, then how do I manage to experience such smoothness? Do I most persistently and clearly experience what is utterly not present before my encountering "eye"? That's madness. It *must* be that something somehow present within the extremities of my act of encountering is engaging in smooth motion and is presenting me with an encounter of that characteristic of its motion. It's just that the something in question is not any of the sense images.

Another thought pressed its way into my presence—one which Albert Einstein, too, once mentioned. How many times has the following occurred? You're sitting in a bus awaiting your departure from the depot (If I remember correctly, Einstein used the image of a train and a train station.[1]). Suddenly, to your momentary astonishment, it appears that the depot has begun to move away from the bus. Two or three seconds of reflection, and you realize with a chuckle that the bus is actually the subject of the motion. As did Einstein, I am constrained to ask myself: "*Is* such the case? Which one *is* the subject of the motion? Is it the bus or the depot—the train or the train-station?

We humans certainly do have a strong tendency to project encountered motions away from the center of consciousness and onto a more distant target. Thus, ancient man thought the Sun was the subject of the diurnal motion so familiar to us all. The Catholic cleric, Nicolas Copernicus, finally convinced us to associate that encountered motion with a target much closer to the observer's center.

> Here, we propose to do the same as Copernicus did in his efforts to account for the celestial motions. Seeing he could go nowhere positing all the heavenly bodies as revolving around the spectator, he took the reverse approach and tried assuming that it is the spectator who revolves and the stars which stand still. We can perform the same kind of experiment upon the immediate apprehension of objects.
> ——**IMMANUEL KANT:** *The Critique Of Pure Reason: Preface To The Second Edition, 1787*, my re-translation—ENH. Compare page 7 of volume 42 of **Great Books Of The Western World** published by William Benton, Chicago, 1952.

Yes! Let us, too, here do somewhat the same thing as Nicolas Copernicus did. If it cannot be that the direct encountereds engage in a kind of motion characterized by continuousness, can it be that the *not directly* encountered encounterer engages in such a motion? Can it be that, like the man in the bus, the encounterer is projecting his motion's continuity outward from his center and onto a more distant target of observation?

It is now fifty years since first I answered the above questions affirmatively in 1957. At the same time, the problem of *anything observable* moving *smoothly* through either time or space required me to go much further and to propose:

AS LONG AS IT'S THE SAME ONE ACT, NO *CONTINUOUS* MOTION CAN BE A MOTION THRU EITHER TIME OR SPACE.

---

[1]On pg. 59 of his book **Relativity, The Special And The General Theory**, Einstein refers to "the illustration we have frequently used of the embankment and the railway carriage". Bonanza Books; New York, undated. © 1961 by the Estate of Albert Einstein.

WHILE THE SAME ONE ACT, ALL _CONTINUOUS_ MOTION IS THE KIND BEST DESCRIBED AS THE "ACTUAL BEING", "ACTUALITY", "PRIMARY ACT" AND "ACT OF EXISTENCE" OF EVERY NON-INHERING SUBJECT.[1]

Such, then, is how I came to believe that, in the prime abstractable, I am encountering something of the very encounterableness of my own immortal soul. Let St. Augustine and René Descartes tell us how the capacity to err or to think proves that one exists. I take a very different tack and say:

I AM AWARE OF CONTINUITY; THEREFORE, WHATEVER I AM, IT IS EXCEEDINGLY UNLIKE ANY AND ALL OF THE THINGS MY SENSES EXPERIENCE.

The prime abstractable is somehow a projection of a characteristic of my own self performing an act so unusual, it could only be an act of _existence_ and, thus, a _primary_ act. That act's smoothness becomes attached to the direct encountereds by means of a kind of optical illusion. It's just that, in this case, the eye seeing things backwards is in no way similar to any _directly_ encountered eye. The encounterer has to be radically different from the direct encountereds, and his motions (_i.e.:_ his _primary_ acts) must be without any kind of change thru time and space. Only his _secondary_ acts (_i.e.:_ intermittent, instantaneous changes from primary act to primary act) involve changes thru time and space. Were such not the case, I would again be faced with the problem of explaining how smoothness could be a characteristic of his motions.

More importantly, the way in which the encounterer projects the smoothness of his _primary_ acts of _existence_ is exceedingly different from what happens to the occupants of a bus suddenly moving from the depot. How so?! Be patient. The answer shall be given shortly.

H: THE ONLY _EMPIRICALLY_ CONFIRMABLE CONCLUSION: THE NECESSARY STRUCTURE OF EVERY ACT OF CONSCIOUSNESS IS THE NECESSARY STRUCTURE OF EVERY BASIC REALITY:

For the moment, the sum total of what we have discovered thus far is this: In the strictest sense of "immediate", I immediately experience (_i.e.:_ encounter, come face to face with, and gaze upon) nothing but my own experiencing self and its states (_i.e.:_ what's wholly and entirely interior to my self). But, whether remembering a past state or pondering a briefly present one, it is glaringly obvious to me that I am face to face with something whose inescapable, absolutely necessary, _a_

---

[1]According to Esoptrics, a non-inhering subject is one which does not—as does the shape of something—need to be intimately bonded to another in order to avoid annihilation. It's a subject able to _stand apart_. In other words, it does not—to avoid annihilation—need to have its extremities inextricably bonded to another's. That, of course is what the verb "to exist" means when taken in the strictest, truest, primary, most definite, and unqualified sense suggested by its etymology.

priori structure is that of a <u>tri</u>ousious <u>object</u>, which is to say that of a whole which is undivided within itself but divided from all others; and yet, its extremities enclose 3 <u>subjects</u> inextricably bonded together. Rather than being a *part* of their same, one *object*, each is so separate in identity, it must be classed as a *subject* and *co-tenant* included in their same one whole. That's so because of how radically different each is from the other 2 from the standpoint of their internal characteristics.

First, there are the direct encountereds and the encounterer. The latter is so different from the former, the two can only be described as two kinds of "stuff", "substrata", "subjects", and "separate identities" so different from one another as to fall into radically different categories even from the standpoint of their very kind of reality and kind of Being.

Next, once it is suggested that the prime abstractable, too, is some kind of reality, then—*on the face of it!*—it is so different from the encounterer and the direct encountereds that it must be a *third* category of reality and Being which is radically different from the other two.

This bit—of three radically different kinds of reality inextricably bonded together within the confines of the same one whole—is the absolutely unavoidable structure of my every act of consciousness. After all, by the very nature of consciousness, there can be no such thing as an act of consciousness save where there are: (1) that which is being aware, (2) that of which the former is being aware, and (3) the act of consciousness joining the two. But, there is no such thing as awareness or observation or confirmation by observation save by means of an act of consciousness; and so, I can never observe or be aware of anything save that which is one of the three subjects present within the confines of my acts of awareness.

Therefore, if there is such a thing as that which is *not* one of the three subjects within the confines of the same, one object, then I cannot confirm the existence of such a reality *by observation*. After all, the only way I can *observe* it is to make it one of the three subjects within the confines of the same one object. On the face of it, that would make such a sought after exception the very opposite of what I was trying, by observation, to prove it to be. In short, it is absolutely impossible for me to prove *anything* by *observation* save the universality of the triousious structure I observe in my every act of consciousness. That being the case, no conclusion is reasonable or in conformity with what I incessantly observe save this:

THE NECESSARY, TRIOUSIOUS STRUCTURE
OF EVERY ACT OF CONSCIOUSNESS
IS THE NECESSARY STRUCTURE OF EVERY BASIC REALITY.

How, then, does such a structure manage to be the inescapable one of every basic reality? By the term "basic reality", I mean, of course, a reality which cannot in *fact* (as opposed to in *fancy, imagination, speculation, theory,* or the like) be broken up either into smaller pieces (**ex. gr.:** as a boulder can be smashed into pebbles) or into more basic realities (**ex. gr.:** as a molecule can be reduced to 2 or more atoms, or an atom reduced to leptons and quarks).

PART II: THE SOURCE OF

EVERY BASIC REALITY'S TRIOUSIOUS STRUCTURE:

I: THE ACTIVITY OF A MIRROR:

Long ago, I reflected upon the prime abstractable and asked myself: "How can that which is not activity be conscious of activity? How does the encounterer encounter a property of himself which—if the encounterer is something other than activity—must be radically unlike himself? How can any encounterer encounter an act of encountering unless the encounterer himself can be described as an act of encountering?" Thus was I driven to conclude this: That which actually moves is itself motion or, as I *prefer* to express it, actuality is what acts.

"But," my mind complained, "how can motion be what moves? To have motion, you must have something moving." I answered myself: "The only way you can grasp the answer to that question, is to encounter *real* motion *itself*. Come face to face with the internal characteristics of *real* motion, and you will see how it is possible that motion is, indeed, a real thing capable of being what is moving."[1]

From that realization, it followed that the prime abstractable is by no means merely a projection of a characteristic of a kind of activity *associated with* the encounterer. Utterly to the contrary, it is a projection of the very encounterableness of the encounterer. In other words, instead of being a projection of something the encounterer is merely **_doing_**, it is a projection of something which the encounterer is also **_being_**. In the encounterer's case, to **_do_** a kind of motion and activity characterized by smoothness and continuousness is to **_be_** a kind of "Being", "act", and "actuality" characterized by smoothness and continuousness.

How does that projection take place? Is it a thing or an illusion? What is an illusion? Is it a nothingness? My mind once suggested: "You're merely noting the difference between what was and what is." I replied: "How can that be? What *was*,

---

[1] In Book One of his **Summa Contra Gentiles**, Chap. 13, Par. 10, St. Thomas Aquinas contrasts how Plato and Aristotle use the word "motion". Plato, he says, uses it to refer to *any* given operation—even mental acts. For Aristotle, says St. Thomas, "motion" refers only to the kind of operation which "belongs only to divisible bodies" (**Vd.:** pg. 88 of Vol. 1 of the University of Notre Dame Press edition of 1975). In other words, for Aristotle, "motion" denotes those changes thru time and space our sense images exhibit, and that's mainly locomotion. Needless to say, *such* "motion" cannot be that which moves. For "motion" to signify what moves, it must—somewhat as it does for Plato—signify an *operation* and, at that, one of a very unusual kind—an operation of a kind so unusual, it can be that which moves. In Book III of his **De Anima**, Chap. 7 (431a 5), Aristotle says *motion* is different from *activity*, since, "in the unqualified sense", activity belongs to "what has been perfected" (**Vd.:** pg. 663 right of Vol. 19 of the **Great Books Of The Western World**). Unlike myself, most people will *insist* "motion" be used the way Aristotle uses it. As a result, some will insist that, in the above, I am, like Bergson and Heraclitus, claiming that change is the only real thing. Well, I am *not* using it Aristotle's way. I'm using it, it seems, the way Aristotle uses "activity". If, then, it confuses you to hear me speak of motion moving, then change it to read "activity is what acts" or "act is what acts" or "actuality is what acts". However difficult it is for carnal minds to do so, one must cease thinking of ultimate realities as little lumps of inert, billiard-ball rigidity and learn to think of them as so inherently dynamic, every iota of their internality is some kind of "whirling dervish" whose whirl, nevertheless, undergoes no kind of change thru either time or space, save when, occasionally, the dervish suddenly stops whirling with one discrete level of powerfulness and starts to whirl with a different discrete level.

is gone. You're telling me I'm aware of the hole in the doughnut without the presence of the doughnut."

Long ago, an almost unbelievable answer came to me: The encounterer is an activity best described as the activity of a mirror. By simply being himself, the encounterer thus engages in the act of trying to mirror something, and the thing which he tries to mirror is himself. He's trying to mirror himself in an attempt to come face to face with himself. But, he does not mirror himself with infinite efficiency. As a result, he gets an inside-out picture of himself and, thereby, comes face to face with an encounterableness *radically* different from the one he set out to encounter. In other words, imperfectly mirroring the activity which he is, he begets a distorted, reversed, negated image of himself—provided that by "negated" you mean "made negative" rather than "annihilated".

But, what is *having* and *exhibiting* this reversed encounterableness? Clearly, it cannot be that the same one, unencountered encounterer—without being two different encounter<u>ables</u>—has two radically different encounterable<u>nesses</u>. The answer is inescapable: It must be that, in mirroring himself with less than infinite efficiency, the encounterer somehow produces a second encounterable with three indispensable features: (1) Its encounterableness is radically different from that of the encounterer's actuality, and (2) it is annihilated, unless (a) the encounterer continues to produce it by continuing to mirror himself imperfectly, and (b) it shares at least some of its extremities with those of the encounterer's actuality.

Let the prefix "non" mean "made negative", and we can say this: The prime abstractable is a non-reality, a non-thing, and a gossamer, almost unreal, mirrored image of the encounterer. At the same time, this non-thing is <u>not</u> another reality *completely separate from*, and *independent of*, the encounterer. Somehow it's another "side" of the same encounterer acting as the *positive* reality projecting this *negative* one. In short, the encounterer is—simultaneously, and within the confines of a same one whole sharing none of its extremities with another—two radically different realities. Thus, though this object and whole called the encounterer is to no extent internally divided in fact, it is two radically divergent kinds of substrata inextricably bonded together and each having an encounterableness so different from that of the other, that we *must* make these contrasts: If one be called a reality, then the other must be called an *anti*-reality; if one be called a "Being", then the other must be called an "*Anti*-Being"; and, if one be called substantial, then the other must be called <u>in</u>substantial. One is parasitic and the other not.

So greatly do the encounterer and his anti-self differ, we can refer to them as two different *subjects* (*i.e.:* separate in identity) inextricably coupled within the confines of the same, one whole which, as such, is, itself, to no extent inextricably affixed to another. It's by mirroring himself imperfectly, that the encounterer produces a "polyousious object (*i.e.:* a multi-subject whole)" so profoundly sealing two subjects together, one of them (*viz.:* the anti-self) *inheres* in the other.[1]

---

[1] In Esoptrics, "to inhere" implies a *parasitic* substratum, which is to say a kind of "stuff" which must ever remain intimately bonded to (*i.e.:* share at least some of its extremities with) a non-parasitic substratum to avoid annihilation. Naturally, a non-parasitic substratum is a kind of "stuff" which does *not need* to be intimately bonded to another in order to avoid annihilation.

ESOPTRICS: LOGIC OF THE MIRROR

If the encounterer is a *mirroring* activity, all of existence's fundamental laws may reduce to the way mirrors operate. The great key, then, may be some kind of "Logic Of The Mirror". Using the Greek word for "mirror" I call that logic "Esoptrics". That logic most immediately suggests to me these two postulates:

FIRST POSTULATE

EVERY FINITE ENCOUNTERER IS—AT ONE AND THE SAME INSTANT, AND WITHIN THE CONFINES OF THE SAME ONE WHOLE—BOTH HIMSELF AND HIMSELF NEGATED IN AN ATTEMPT TO ENCOUNTER HIMSELF BY MIRRORING HIMSELF.

SECOND POSTULATE

IN THE REALM OF FINITE NON-INHERING REALITIES, MIRRORED IMAGES ARE NOT EQUALLY REAL AND, THEREFORE, ARE RADICALLY DIFFERENT FROM ONE ANOTHER FROM THE STANDPOINT OF THEIR ENCOUNTERABLENESS.[1]

J: THE INFINITE VS. THE FINITE MIRROR:

The preceding merely recasts one of St. Thomas Aquinas's most famous principles: no *creature's* essence (*i.e.: what* it is) is the same as its existence (*i.e.: the means by which* it is what it is). As for *what* an encounterer is, he is what he encounters—mainly the prime abstractable and, secondarily, the direct encountereds and encounterables. As for the *means* by which he is what he is, he is the activity of a mirror. Hopefully, in spelling out the relationship between the prime abstractable and the encounterer, I've made it clearer how such disparity could occur. Let's give a figurative explanation of what I mean by an *infinite* encounter, and maybe the picture will become even more intelligible.

An infinite encounter is one involving an infinite encounterer. An infinite encounterer is one who engages in the act of encountering at infinite velocity. His act of mirroring himself—in an attempt to encounter himself—involves the *full* development of all his potency (Say "energy", if you prefer.) and is, consequently, a *pure act*. In a manner of speaking, he emits a beam of light traveling at infinite velocity. If you do not mind analogies cast in such self-contradictory language, we can express the result like this: The beam of light covers an infinite distance in an infinitely small period of time. That means it *circles* infinity *instantly*, comes back to itself unchanged (*i.e.:* it is still "pointed in the same direction".), and encounters itself in the very act of being emitted. Such an occurrence can be *symbolically* expressed in the next page's image where "E" represents the infinite encounterer emitting the beam, and "I" represents the opposite "ends" of infinity. The arrow to the right of E signifies the beam being emitted, and the arrow to the left of E signifies the beam being received.

---

[1] For those familiar with the physicists' law of parity and its overthrow, the above may ring a bell.

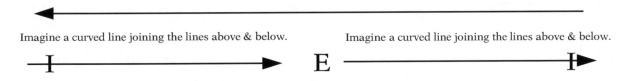

Imagine a curved line joining the lines above & below.   Imagine a curved line joining the lines above & below.

DRAWING #22: The Infinite Encounter.

Thus, the *infinite* encounterer encounters a positive, non-reversed, non-negated "picture" of Himself. He thus encounters the internality of His act of encountering exactly as that act's internality is. In other words, by His fundamental activity, He causes Himself to observe no internal characteristics but those of His own fundamental activity itself. Oppositely, the *finite* encounterer—by his most fundamental activity (*i.e.:* his "primary act" and "act of Being")—causes himself to have *observed* internal characteristics radically different from the *un*observed internal characteristics of his fundamental activity itself. As a result, the essence of the *finite* encounterer is purely a mask continually projected by his face; whereas, the essence of the *in*finite encounterer is forever His face itself. We can describe the latter result as:

ENCOUNTERING IS ENCOUNTERING ENCOUNTERING.

That means encountering is *both* the encounterer, *and* the encountered, *and* the act uniting the two. You might say we have one act but three activities, or one act but three functions. In other words, at one and the same instant, and within and throughout the confines of the same, one act, the *pure* act of *infinite* encountering produces, is, and stands within, three frames of reference.

The remarks I have made about the First Encounter allow us to formulate what I call the third and fourth postulates of Esoptrics. I state them thusly:

THIRD POSTULATE

IN THE REALM OF THE INFINITE ENCOUNTER, EVERY ENCOUNTERER IS—AT ONE AND THE SAME INSTANT, AND WITHIN THE CONFINES OF THE SAME, ONE, WHOLE—BOTH HIMSELF, HIMSELF MIRRORED (THOUGH NOT NEGATED), AND THE ACT UNITING THE TWO.

FOURTH POSTULATE

IN THE REALM OF THE INFINITE ENCOUNTER, MIRRORED IMAGES ARE EQUALLY REAL AND, THEREFORE, IDENTICAL FROM THE STANDPOINT OF THEIR ENCOUNTERABLENESS.

Contrast the preceding with the *finite* encounterer. Failing to develop all his potentialities, he emits a beam of light at less than infinite speed. For that reason, the beam simply travels away from the encounterer forever, and the act is eternally incomplete, unless something intervenes to boomerang it back to its source. For that reason, we must hypothesize the existence and intervention of the First Encounter. It is He Who boomerangs the beam back to its source. But, as the beam boomerangs back, it is turned around and reversed. That is to say, when it is reflected back by the First Mirror (*i.e.:* God), what comes back is not equally real with regard to what went out. That can be expressed in the following *two-stage* figure (Note that the *infinite* encounter involves only *one* stage. That's very important.). The first stage, of course, signifies the *emission* of the light beam, and the second stage signifies the *reception* of the light beam.

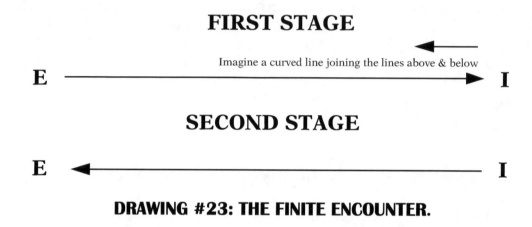

FIRST STAGE

Imagine a curved line joining the lines above & below

E ⟶ I

SECOND STAGE

E ⟵ I

DRAWING #23: THE FINITE ENCOUNTER.

Thus, when the beam comes back to its source, it is negative, and the encounterer encounters a mere ghost of himself. That's why real motion appears to us as nothing more than a gossamer, empty smoothness underlying the encountered transitions (*i.e.: secondary* acts) of the direct encountereds. That's why the *reflexively* encountered *in*substantial essence of motion is unimaginably different from the *substantial* encounterableness it has in the encounterer's actuality. As St. Thomas Aquinas might express it: The essence which the finite E has in E's own eyes is radically different from the essence which E has as the activity of a mirror; and so, essence and existence are not the same for those whose mirroring activity is merely finite.

With all the above thoughts racing through my perhaps crazy head, I next said to myself: "It seems to me that the encounterer can be described as a thesis seeking an anti-thesis. In this anti-thesis, he hopes to encounter himself and to achieve a synthesis in which he will then—in the ecstasy of *total* self-discovery—enjoy quiescence from his chief yearning which is: to come face to face with that *actual* version of himself which is most properly called either 'an act of existence' or 'an act of encountering'." Therefore, every encounterer—whether God, an angel or a human soul—is a triune mirroring activity comprising a positive and a negative element seeking neutrality as the third element. In God and every occupant of infinity, that neutrality is achieved. In humans and every occupant of the finite world, that neutrality is not achieved.

As early as 1957, and certainly no later than 1961 (Writings of mine from that year document it.), I turned to Algebra's symbolic logic to produce the following expression of The Infinite Encounter:

$$(+A) \otimes (-A) \longrightarrow 0$$

Drawing #24: Algebraic Expression Of The Infinite Encounter.

When I first produced this drawing, I looked upon the (+A), the (-A) and the zero as the three elements of the encounter. It wasn't until 1994 that I realized the circled X is the third element, and the zero merely expresses what (+A) and (-A) achieved by means of the circled X. It then became the circled X, and not the zero, which represents the act of encountering whereby (+A) comes face to face with its actual self in (-A). More precisely, it is ⊛ whereby Encountering comes face to face with itself as three distinct functions of the same, one instance of the same, one set of internal characteristics.[1]

In finite encounters, the all-important neutrality is not always achieved; and so, the fundamental algebraic expression of the finite encounter must be:

$$(+a) \circledast (-a) \longrightarrow ?$$

Drawing #25: The Algebraic Expression Of The Finite Encounter.

Drawing #24 is the infinite and #25 the finite version of what can be called the principle of a mirroring act of the first order. They can also be called the infinite and finite versions of the principle of principles or the first and supreme principle of Esoptrics. That's because, in either case, it's the ultimate, supreme definition of all reality, and contains within itself all the diversification possible to reality. In its infinite version, it is the definition of the *logical* essence of God and, as such, is the ultimate ground of all that which is intrinsically possible.

---

[1] Some will accuse me of here trying to arrive at the mystery of The Blessed Trinity by means of human reason. They are sadly mistaken. There is nothing in Esoptrics to lead to the conclusion that each of the three functions is a *person*. The idea of Three Divine Persons in The One God is very much *in line* with Esoptrics' basic principles, but is by no means a *necessary, logical conclusion* of those principles, and, as St. Thomas Aquinas states (*Vd.: Summa Theologica: Pars Prima*, Q32, A1, Resp.), it is the trinity of *persons* which cannot be known by reason. Even for Esoptrics, the trinity of *persons* remains something it cannot explain.

We may as well state, without further ado, that the ultimate ground or foundation of all possibility is found in the Supreme Being, in the nature of God.

. . . .

God's essence is the ultimate ground of intrinsic possibility.
——**FR. CELESTINE N. BITTLE, O.M. CAP.:** *The Domain Of Being—Ontology* as given on pages 70 and 77 published by Bruce Publishing Co., Milwaukee, 1942.

If there is to be any multiplicity in reality, it can only come about by applying the first principle to itself. That is to say, we must modify the first principle by itself. But, **to be** is to be involved in a mirroring activity, and the first principle is the definition of that mirroring activity. Therefore, to multiply it, we must mirror it. Obviously, mirroring it will produce a configuration of *two* such principles. One will be "inside" the mirror, and the other will be "outside" the mirror, and they will be reverse images of one another. Such will give us the principle of a mirroring act of the second order. It can be called the secondary form of seeking synthesis or the secondary form of seeking neutrality. We will generally refer to it, though, as: "the second principle of Esoptrics". Using Algebra's symbolic logic, it can be stated so:

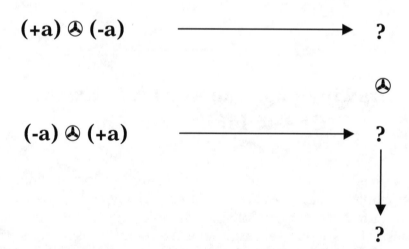

DRAWING #26: ALGEBRAIC EXPRESSION OF ESOPTRICS' SECOND PRINCIPLE.

As said, drawings #25 and #26 give us an *algebraic* expression of Esoptrics' first and second principles. Those two principles can also be given a *geometrical* expression. In that case, though, we must bear in mind what I've said repeatedly—namely: Every geometrical expression of Esoptrics' principles is merely an allegory presenting us with a purely *figurative* expression of those principles, and the reason for resorting to such pictures is to make it easy for the carnal minds of sensation lovers to grasp what would otherwise be too abstract for them. The use of Geometry's pictures is far from being an attempt to show what ultimates "look like". Here, then, are two geometrical attempts to make drawings #25 and #26 more easily intelligible.

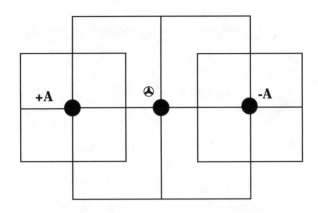

DRAWING #27: GEOMETRICAL EXPRESSION OF ESOPTRICS' FIRST PRINCIPLE.

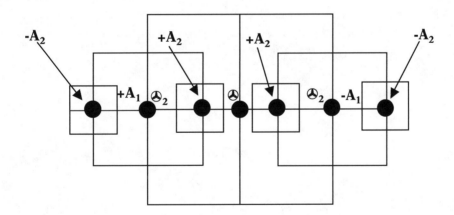

DRAWING #28: GEOMETRICAL EXPRESSION OF ESOPTRICS' SECOND PRINCIPLE.

Bear well in mind these 2 geometrical expression and their algebraic counterparts. They are the keys to what Esoptrics says are: (1) the overall eightfold structure of the universe, and (2) the basic eightfold structure of the hydrogen atom. Drawing #27 shows only the first 2 of 8 levels, and drawing #28 shows only the first 3. At the first level, there's only the *core* ultimate (*i.e.:* ☻). At the 2nd level, there are the 2 *subordinate* ultimates subordinate to the *core* level. At the 3rd level, there are 4 subordinates using the first 2 subordinates as their cores. At the 4th level, there will be 8 ultimates; at the 5th level, 16 ultimates; at the 6th level, 32 ultimates; at the 7th level, 64; and, at the 8th, 128.

PART III: THE SEXPARTITE NATURE OF INFINITY & HOW IT IMPARTS TO EVERY FORM A SEXPARTITE POTENCY:

K. PRELIMINARY CONCEPTUALIZATION:

We've said the necessary, triousious structure of every act of consciousness is the necessary structure of every basic reality. Assuming that to be true, we can formulate what we can perhaps call the fifth postulate of Esoptrics:

THE FIFTH POSTULATE

EVERY BASIC REALITY IS A NON-INHERING WHOLE[1] AND A POLYOUSIOUS OBJECT ENCASING THREE SUBJECTS—TWO OF WHICH ARE DIRECT OPPOSITES, THE THIRD OF WHICH IS A BRIDGE BETWEEN THE OTHER TWO.

Can one imagine something which would fill that bill? Yes! Turn to the idea of *potentiality* versus *actuality*. By a subject's "actuality," of course, I mean the sum total of all its *current* internal characteristics (*i.e.* all it *is being inside itself*); by its "potentiality," I mean the sum total of all its *future* internal characteristics (*i.e.* all it *can become inside itself*). That distinction now allows us to turn to the concept allegorically depicted in the geometrical picture below in drawing #29.

DRAWING #29: GEOMETRICAL EXPRESSION OF THE PRELIMINARY CONCEPT OF THE SIMPLEST POLYOUSIOUS OBJECT.

The small, solid black circle at the center represents one of the co-tenants in a polyousious object. The much larger, transparent circle surrounding the solid black one represents another one of the co-tenants in that polyousious object. The former represents the *actuality* of a non-inhering subject. The latter represents the *potentiality* of that non-inhering subject and is, itself, a subject—an *inhering* one.

---

[1] Remember: A non-inhering whole is one which does not *need* to be intimately bonded to another to avoid annihilation, which is to say it does not *need* to share it extremities with another's.

Double check what I just said: The subject's potentiality is by no means a mere mental abstraction. However transparent and devoid of *palpable* characteristics, it does have an encounterableness and is a *subject*, which is to say separate in identity and real—so real it shrouds the non-inhering subject whose potentiality it is.

Nevertheless, it is *not equally* as *real* as is the non-inhering subject. After all, if the potentiality of a given subject be disjoined from that subject, that potentiality is nothing, which is to say, if you attempt to dissolve its bond with its subject, you will annihilate it. Should you do that, the *non*-inhering subject—cut off from its potentiality—would *not* be *annihilated*, but could undergo no change whatsoever and would become eternally and totally immobile as in eternally damned.

So far, then, we have an object comprised of a non-inhering subject and an inhering one. Each is truly separate in identity (*i.e.:* a subject), because the potentiality of a non-inhering subject—though it *shall become* that subject—is *not yet* that subject. On the other hand, the potentiality of a *given* non-inhering subject cannot be separated from *that given* subject; otherwise, it is not the potentiality of anything and, therefore, is not anything at all.

We, though, are supposed to be devising a polyousious object having **three** co-tenants? Where is the *third* subject?

Imagine the diameter of the small circle is 1 inch, and the diameter of the larger circle is 10 inches. In that case, compare the whole object's diameter to the small circle's diameter, and the ratio is 10 to 1. That ratio tells us that the non-inhering subject is, for now, "actualizing" one-tenth of its full potential. As some might prefer to express it, the encounterer is currently converting only a tenth of its available energy into work. To put it still another way, the "engine" is currently employing only one-tenth of its available horsepower.

Let us imagine that, as long as that ratio endures, the non-inhering subject can be described as continuously being and performing the same, one, changeless, primary act. As it is doing so, it is acting in the presence of another subject which is both very different from itself and bonded to it. Shouldn't that have some kind of effect? I suggest it should. I suggest such a situation produces—between the two subjects—what we can call a "drag inducing differential" of a definite quantity. Naturally, that quantity is determined by the 10 to 1 ratio.

That "drag inducing differential" will produce a *line* of *force* (Say "line of tension" or "vibration" or "friction" or "stress", if you prefer.) extended between two terminals. One end of the line is the *non*-inhering subject; the other end is the *inhering* subject. Furthermore, every last one of that line's internal characteristics will be determined by that 10 to 1 ratio.

Imagine that, after the 10 to 1 ratio has endured for a while, the non-inhering subject suddenly changes instantaneously to a diameter of 2 inches. It does so by instantaneously changing the quantity of its *potentiality* which is now *actuality* instead. In other words, in a durationless instant, the engine suddenly switches from employing *one*-tenth of its horsepower to employing *two*-tenths.

Because the quantity of its actuality is now *changed*, the non-inhering subject is now performing a different act. Since the diameter of the whole object is still 10 inches, the 10 to 1 ratio has changed to 5 to 1 (*i.e.* 10/2 = 5/1). Do you see what's coming? The old line of force has been annihilated. It has ceased to be, and, because

139

the former drag inducing differential has been replaced by one of a different quantity, a different line of force is now being generated by that differential.

Do we now have a polyousious object embracing three *subjects*? We most certainly do in at least one sense—namely: Actuality, potentiality, and a line of force are each a kind of *Being* and substratum so different from the other two, that no one of them could ever be construed as either a part or an *internal* characteristic of either one of the other two. Each has an encounterableness so unique in comparison to the other two, that there is simply no way you can—*not to any extent whatsoever!*—intermingle one with either of the other two. On the other hand, the potentiality is the potentiality **of** the actuality, and the line of force is the product **of** the other two (Shades of the *Filioque*!). In short, each of the three is an encounterable, a kind of *Being*, and a kind of substratum so different from the other two in internal characteristics and encounterableness, that it is about as "separate in identity" as it can be; and yet, the potentiality and the lines of force are so totally **of** the actuality, they are certainly not as separate in identity as different people are. Hopefully, that will not prove *too* confusing to you.

At this point, take special note of some crucial differences between the three co-tenants of our polyousious object: The *non*-inhering subject itself is activity itself, impermeable, and endures indefinitely no matter what happens. The *inhering* subject (*i.e:* the potentiality of that first subject) is perfectly passive and permeable, and it endures indefinitely *but* only as long as it remains intimately bonded to (*i.e.:* shares at least some of its extremities with) that first subject. The third co-tenant (*i.e.* the stress) is a mere projection momentarily generated by the interaction between the other two, and it endures for only *one* of the non-inhering subject's acts.

The three co-tenants of our polyousious object are three kinds of *Being*. The center circle is "actual Being", and the outer circle is "potential Being". As for the line of force generated by the interaction between actual and potential Being, it is perhaps best called "accidental Being". However, I've not really made up my mind on that issue for reasons not worth discussing.

L. PRELIMINARY GEOMETRICAL REPRESENTATION:

We now have a very unusual concept—namely: one whose purely *logical* content can be expressed by a *picture*. Still, the result could hardly be called a mathematically precise theory of the universe. Obviously, our concept and "picture" of a polyousious object must expand dramatically.

Drawing #30 on the next page—if looked at closely—presses upon us an obvious question: Is there a non-inhering subject with *unlimited* potentiality? But, if its actuality is less than its potentiality, what would be the ratio of the whole to its actuality? Wouldn't it be $10^{infinite}$ to $10^{-infinite}$? Since that strikes me as mathematical nonsense, the ratio must be 1 to 1, which is to say all the potentiality has been converted into actuality, and the whole object is actuality and actuality alone.

In case what I just said above is not clear, consider this example: A group sets out to raise $1,000,000.00. As they begin, the ratio of money raised to money wanted is obviously: 0/$1,000,000.00. When they finish raising all the money, the ratio of money raised to money wanted will be $1,000,000.00/$1,000,000.00 = 1/1. On the other hand, the ratio of money raised to money *left* to be raised is $1,000,000.00/0.

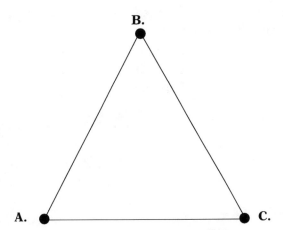

DRAWING #30: PRELIMINARY GEOMETRICAL EXPRESSION OF THE FIRST POLYOUSIOUS OBJECT.

Is the infinite, non-inhering subject a polyousious object? How *could* it be? And yet it *must* be, *if* the postulates and principles of Esoptrics are correct. Clinging to that insight, I cannot escape the conclusion: It must be the infinite, non-inhering subject is The First *Polyousious Object* and, therefore, somehow **is**, and occupies, three frames of reference each of which—though equal to, and equally related to, the other two—is somehow distinct from the other two.

That brings us to drawing #30. The points A, B, and C represent the three frames (or "points," if you prefer) of reference encompassed by The First Polyousious Object, and I have drawn them as corners of a triangle in order to express the fact each is equally related to the other two.

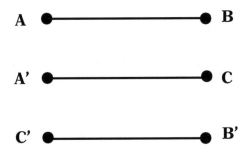

DRAWING #31: ADVANCED GEOMETRICAL EXPRESSION OF THE FIRST POLYOUSIOUS OBJECT.

Examining the triangle, I am inexorably drawn to another insight: I can look at any one of the three points in *two* different ways. For example, A can be considered either as A related to B or as A related to C. Can it be? Is *"A related to B"* somehow distinct from *"A related to C"*? In that case, we can *double* the number of reference points and say: A = A related to B; A' (*i.e.* A *prime*) = A related to C; B = B related to A; B' (*i.e.* B *prime*) = B related to C; C = C related to A; and C' (*i.e.* C *prime*) = C related to B. As you can see in drawing #31 on the prior page, it's merely a case of considering the three arms of the triangle in isolation from one another.

The instant we isolate the three arms that way, another insight thrusts itself upon us: The potentiality of every polyousious object encompasses six potencies which divide into three *sets* each comprised of two potencies. In each set, each of the two potencies is the opposite of the other. Thus, A and B are mutual opposites within the first set; A' and C are mutual opposites within the second set; and C' and B' are mutual opposites within the third set. In keeping with what we've already said, each of these three sets should be equally related to the other two.

It should take little effort to see that the previous paragraph's concepts can be expressed pictorially by drawing three axes each of which is *perpendicular* to the other two. In other words, by making each perpendicular to the other two, we express the fact that each is equally related to the other two. Such thoughts lead to drawing #32 stated so:

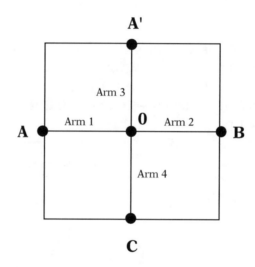

DRAWING #32: ACTUALITY & SIX POTENCIES

Looking at drawing 32 above, please note one factor first: As you may have already perceived, I am confronted therein with a rather formidable problem: I cannot draw three-dimensional figures on a flat surface. In other words, I need to take the *three* lines in drawing #31 and create a six-pointed **star** by making each of the three lines perpendicular to one another. It's easy enough to incorporate *two* arms into a **cross**, as I've done in drawing #32. But, how do I add in the *third* arm—one which needs to be perpendicular to the front and rear surfaces of the sheet of paper?! *I* certainly have no idea how to do it in a way which won't cause confusion. That being the case, I must implore the reader to *imagine* the third arm piercing the page at the center of the cube and, thereby, putting C' out in the air *behind* the page and B' in the air in *front* (or, if the page is lying flat on a table, C' is in the air *below* the page's rear and B' in the air *above* the page's front). Drawing #32 will then be a *cube* rather than a *square*.[1]

As you use your mind to add a third axis to the above figure, be sure to re-mind yourself again of a crucial qualification I've repeated *ad nauseam*: Drawing #32 is merely a pictorial expression of the purely *logical* principles constituting every basic reality, and it describes those principles solely in the terms of a geometrical expression of how they relate to one another. Drawing #32 is not an attempt to give you an encounter of the encounterableness of a basic reality. In other words, drawing #32 is not a "photocopy" of what you would "see" were you to come face to face with the internality of an ultimate. From God's point of view, the points A, B, C, A', B', C', and 0, are *not* outside of one another **in space**; they are outside of one another **in logical principle** and in logical principle **alone**. Drawing #32 depicts the points as outside of one another in space because that's the simplest and most intelligible way for me to go about communicating the logical principles to you.

If drawing #32 applies to God, then 0 (*i.e.:* the actuality of the cube) is coterminous with the rest of the cube, and the outer perimeter of the cube represents the "limits" of *infinite* potentiality converted into *infinite* actuality. If drawing #32 applies to a creature, then 0 may or may not be coterminous with the rest of the cube, and the outer perimeter of the cube represents the limits of whatever potentiality God has made available to His creatures.

Such, then, are the lines of thinking which lead to Esoptrics and its conclusion that The Infinite is an extra-cosmic influence which—by means of the six points of reference intrinsic to The Infinite—imparts a six-fold potency to every form created by The Infinite. Esoptrics refers to these six points of reference as a "heavenly sextet". Subsequent documents shall present us with far more than one heavenly sextet and then, by presenting each sextet's members as shifting roles, shall explain how forms have definite patterns of rotation with definite ratios between the way they rotate around each of their three axes.

PART IV: THE MATHEMATICS OF A MIRROR:

M. FOUR PROGRESSIONS DEALING WITH THE NUMBER 2:

The idea of a mirror suggests several progressions based upon the number 2 and the powers of 2. Let's now examine those progressions.

Imagine that you are standing near a mirror. Hold your right hand up in front of the mirror. For the sake of comfort, face the palm inward toward the mirror. One will now have two hands—one *in* the mirror and one *outside* of the mirror. Furthermore, the two hands will be reverse images of one another. For, the hand outside of the mirror is a right hand, and its thumb is on the left side of the hand, whereas the hand inside of the mirror is a left hand, and its thumb is on the right side of the hand. Suppose we could reach into the mirror, could take hold of the

---

[1]The cross and the six pointed star of David are probably the two most important symbols in the Judeo-Christian tradition. Do you suppose . . . ?! No! It couldn't be! . . . . Could it?!

hand in the mirror, and could bring it out to our side of the mirror. We would then have two real hands which are reverse images of one another.

Now hold those two hands up in front of the mirror. We will now have two sets each containing two hands, and the two sets will be reverse images of one another. Remove the set in the mirror; add it to the set outside of the mirror; and that will give us a set composed of four hands.

Hold that set of four hands up in front of the mirror, and we will now have two sets each containing four hands. Obviously, if we keep that up, we are merely progressing by the powers of 2, and each new set in the mirror has an internal order which is the opposite of the order found among all the sets which preceded it.

Consider now this symbol: "1^{1m}". It means: "one to the first mirror". That is to say, it is the number 1 followed by an exponent telling us to raise the number 1 to a particular power of 1, and the power to which we must raise the number 1 is described as: "1 mirrored once". Now then, any one object mirrored once produces two such objects. Thus, $1^{1m} = 2$, which is the same as 2^1. Next, consider this symbol: "1^{2m}". It means: "1 to the second mirror". It tells us we must raise the number 1 to a power described as: "1 mirrored twice". Now then, any object mirrored once produces two such objects, and if you then mirror those two, you will have four such objects. Thus, $1^{2m} = 2^2 = 4$. Continuing such a line of reasoning, we can make these equations:

$$1^{3m} = 2^3 = 8$$
$$1^{4m} = 2^4 = 16$$
$$1^{5m} = 2^5 = 32$$
$$\text{etc.}$$

To make things a bit more complicated, consider this symbol: "1^{m1}". You will note that—in the exponent—the relationship between the letter and the number is reversed. What does such a symbol mean? Well, it means this: "1 to the first mirror of 1". Thus, the exponent tells us to raise the number 1 to the first mirror of 1, in order to get the exponent of the number we're examining. Thus, $1^{m1} = 1^{2m} = 4$. Suppose, though, the symbol reads thusly: 1^{2m1}. In that case, we must take the second mirror of one in order to establish the exponent of the number, 1. Thus, $1^{2m1} = 1^{4m} = 2^4 = 16$. Perhaps the reader can now grasp the following symbols.

$$
\begin{aligned}
1^{1m1} &= 1^{m1} &= 1^{2m} &= 2^2 &= 4. \\
1^{2m1} &= 1^{m2} &= 1^{4m} &= 2^4 &= 16. \\
1^{3m1} &= 1^{m4} &= 1^{8m} &= 2^8 &= 256. \\
1^{4m1} &= 1^{m8} &= 1^{16m} &= 2^{16} &= 65{,}536. \\
1^{5m1} &= 1^{m16} &= 1^{32m} &= 2^{32} &= 4{,}294{,}967{,}296.
\end{aligned}
$$

We have now established two of the progressions with which the logic of the mirror deals. There are two other very important ones. Perhaps I should say that I myself have currently mastered only two more of them. Let us proceed to establish the remaining two.

Take an object and set it to the side. Take another object exactly like it, and mirror it. Take the image from the mirror; couple it to the one in your hand; and set the two next to the first single object. That now gives you three objects (*i.e.:* $1 + 1^{1m}$). Take another object exactly like the original object, and mirror it. Take the image from the mirror; couple it to the one in your hand; and set the two next to the other three. That now gives you five objects (*i.e.:* $1 + 1^{1m} + 1^{1m}$). Continuing that process gives you another important progression of Esoptrics.

We now have the following three progressions. I should not have to point out, that—in each progression—the lower lines are each another way of writing what is written in the lines above it.

I. THE GREAT PROGRESSION:

$$
\begin{aligned}
1 &\quad + 1^{1m} &&+ 1^{m1} &&+ 1^{2m1} &&+ 1^{3m1} &&+ 1^{4m1} &&+ 1^{5m1} &&\text{etc.} = \\
1 &\quad + 2^{1} &&+ 1^{2m} &&+ 1^{m2} &&+ 1^{m4} &&+ 1^{m8} &&+ 1^{m16} &&\text{etc.} = \\
1 &\quad + 2^{1} &&+ 2^{2} &&+ 1^{4m} &&+ 1^{8m} &&+ 1^{16m} &&+ 1^{32m} &&\text{etc.} = \\
1 &\quad + 2^{1} &&+ 2^{2} &&+ 2^{4} &&+ 2^{8} &&+ 2^{16} &&+ 2^{32} &&\text{etc.}
\end{aligned}
$$

II. THE SECONDARY PROGRESSION:

$$
\begin{aligned}
1 &\quad + 1^{1m} &&+ 1^{2m} &&+ 1^{3m} &&+ 1^{4m} &&+ 1^{5m} &&+ 1^{6m} &&\text{etc.} = \\
1 &\quad + 2^{1} &&+ 2^{2} &&+ 2^{3} &&+ 2^{4} &&+ 2^{5} &&+ 2^{6} &&\text{etc.}
\end{aligned}
$$

III. THE BASIC PROGRESSION:

$$
\begin{aligned}
1 &\quad + 1^{1m} &&+ 1^{1m} &&+ 1^{1m} &&+ 1^{1m} &&+ 1^{1m} &&+ 1^{1m} &&\text{etc.} = \\
1 &\quad + 2 &&+ 2 &&+ 2 &&+ 2 &&+ 2 &&+ 2 &&\text{etc.}
\end{aligned}
$$

Between the great and secondary progressions, there is obviously one which should read as follows:

IV. THE INTERMEDIATE PROGRESSION:

$$
\begin{aligned}
1 &\quad + 1^{m1} &&+ 1^{m2} &&+ 1^{m3} &&+ 1^{m4} &&+ 1^{m5} &&+ 1^{m6} &&\text{etc.} = \\
1 &\quad + 2^{2} &&+ 2^{4} &&+ 2^{6} &&+ 2^{8} &&+ 2^{10} &&+ 2^{12} &&\text{etc.}
\end{aligned}
$$

✡✝✡✝✡✝✡✝✡✝✡✝✡✝✡✝✡✝✡✝✡✝✡✝✡✝

The period extending from the beginning of Christian speculation to the time of St. Augustine, inclusive, is known as the Patristic era in philosophy and theology. In general, that era inclined to Platonism and underestimated the importance of Aristotle. The Fathers strove to construct on Platonic principles a system of Christian philosophy. They brought reason to the aid of Revelation. They leaned, however, towards the doctrine of the mystics, and, in ultimate resort, relied more on spiritual intuition than on dialectical proof for the establishment and explanation of the highest truths of philosophy.

——**WILLIAM TURNER:** in his article **Scholasticism** as found near the bottom of the lower left-hand column of page 548 in volume XIII of the 1941 Gilmary Society reprint edition of the 1913 **Catholic Encyclopedia**. (I, of course, lean "towards the doctrine of the mystics" and rely "more on spiritual intuition than on dialectical proof for the establishment and explanation of the highest truths of philosophy." What does that mean? First, let's get clear about "intuition". "Intuition" is often wrongly used to mean an *instinctive premonition* as in feminine intuition. Properly used, "intuition" signifies the *bare* act of noticing what's *intra*-mentally present to one's gaze. It means keen observation uncomplicated by any additional mental operations. To promote humanity's intellectual advancement, Science turns to what Aristotle calls: "keenly focusing upon [intuiting – ENH], what the external world presents to us"; whereas, *dialectical* philosophers, "having made few external observations, quickly limit themselves to abstruse argumentation." Science's method is to apply telescopes, microscopes, thermometers, barometers, etc., to *extra*-mental realities in order to find support for their inferences. *Dialectical* Philosophy's method is what we see in Plato's Socratic dialogues: to find support for one's inferences in nothing but hair-splitting debates over words and the concepts they seek to describe. *Mystical* Philosophy's method is to seek from God an unusually powerful ability to peer into the innermost depths of one's own self far more deeply than humans ever do on their own. For, it hopes that—in such a mind-boggling depth of self-knowledge— "spiritual intuition" shall *intra*-mentally observe eternal *a priori* first principles hidden from even the most brilliant of scientists and dialecticians. Clearly and unequivocally, Esoptrics is the result of such a method and, if correct, stands forever as conclusive proof that the mystics alone were on the right path to those *really* **highest** truths which Science and *dialectical* Philosophy have ever sought in vain. To be honest, though, this lover of mystical Philosophy must admit that, without the aid of scientists and dialectical philosophers like St. Thomas Aquinas, I would have been hopelessly lost in the maze of data presented to me by spiritual intuition's intra-mental gaze. Perhaps: Dialectics = thesis; Science = anti-thesis; Mysticism = synthesis.)

146

Document #7:

CURVILINEAR MOTION:

A. PATTERNS & ANTI-PATTERNS OF DIVINE ROTATION:

On pages 53 thru 55, I gave my reader a fairly detailed explanation of what curvilinear motion is for Esoptrics—namely: It is the result of the fact that each of the 6 sides of each form must maintain its eternal orientation to its one of the 6 heavenly reference _points_ even as its one of those 6 points logically trades roles with its 5 companions according to the heavenly _benchmarks_ (**a/k/a:** heavenly _frames_ of reference). Combined with document #6's explanation of how The Infinite presents each form with 6 points of reference, it should now be clear what stands as the logical basis for such a relationship. What's just been said is illustrated by the following 2 drawings. Just be sure, as usual, to bear in mind that these drawings are merely "Geometry speak" trying to help you grasp the abstract, logical principles of Algebra. Also remember to imagine B' in the air above, and C' in the air underneath, the paper.

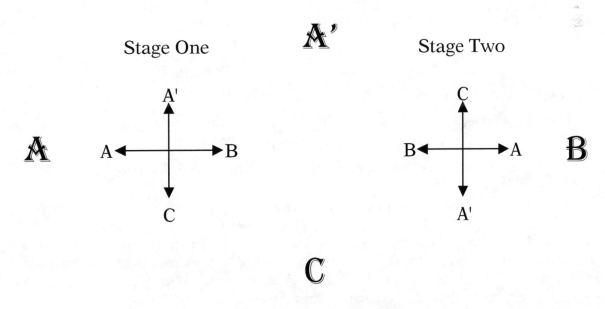

**DRAWING #33: THE TERTIARY PLANE'S
SIMPLEST PATTERN OF DIVINE ROTATION.**

Drawing #33 shows us the pattern of divine rotation which applies solely to a form whose native and current OD is 1. There's no need to represent stage *three*, since it would merely be a repeat of Stage 1 and would be the start of another rotation. In drawing #33, the smaller, plain letters represent the 6 members of the pertinent heavenly sextet, which is to say the 6 heavenly reference *points* pertinent to this particular instance of the form we've previously called Alpha1. The larger, more elaborate letters represent the 6 members of the pertinent heavenly benchmarks, which is to say the 6 heavenly *frames* of reference which apply at least to this particular heavenly sextet. Look close, and you'll come to a very, very crucial realization: It is not possible to tell whether the above pattern is—in the terms of Geometry speak—either clockwise or counter-clockwise.

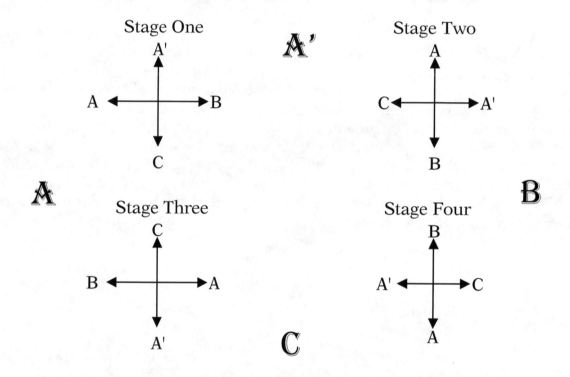

DRAWING #34: THE TERTIARY PLANE'S NEXT TO SIMPLEST PATTERN OF DIVINE ROTATION.

The above shows us the pattern of divine rotation which applies solely to a form whose native and current OD is 2. There's no need to represent stage *five*, since it would merely be a repeat of Stage 1 and would be the start of another rotation. Here again, the smaller, plain letters represent the 6 members of the pertinent heavenly sextet, which is to say the 6 heavenly reference *points* pertinent to this particular instance of the form native to, and currently at, OD2. Here again, the larger more elaborate letters represent the 6 members of the pertinent heavenly benchmarks, which is to say the 6 heavenly *frames* of reference which apply at least to this particular heavenly sextet. Look close, and you'll come to a very, very crucial realization: It is now possible to tell whether the above pattern is—in the terms of Geometry speak—either clockwise or counter-clockwise. Only for forms currently at, and native to, OD1 is it impossible to determine whether the rotation in the tertiary plane is clockwise or counter-clockwise.

As I said, it is Geometry speak which, for economy's sake, speaks of *clockwise* and *counter-clockwise* rotations. If B' is in the air above, and C' in the air underneath the page, Algebra speaks so:

PLANE:	WHAT SUPPLANTS WHAT	AROUND
Primary:	A' → B'; B' → C; C → C'; C' → A'	AB
Secondary:	A → B'; B' → B; B → C'; C' → A	A'C
Tertiary:	A → A'; A' → B; B → C; C → A	B'C'

CHART #3: ROTATIONS IN THE THREE PLANES

PLANE:	WHAT SUPPLANTS WHAT	AROUND
Primary:	A' → C'; C' → C; C → B'; B' → A'	AB
Secondary:	A → C'; C' → B; B → B'; B' → A	A'C
Tertiary:	A → C; C → B; B → A'; A' → A	B'C'

CHART #4 REVERSE (*I.E.:* MIRRORED) ROTATIONS IN THE THREE PLANES.

On page 55, I stated that the number of states per rotation in the tertiary plane = 2 times the *current* OD of the form with which the heavenly sextet is associated. In other words, if S_3 = the number of states (*i.e.:* for Geometry speak: locations, positions, places, etc.) in the tertiary plane and D_c = the current OD of the form with which the heavenly sextet is associated, then:

Formula #7: $S_3 = 2D_c$

In drawing #33, we see that formula illustrated in the case of a form native to, and currently at, OD1. 2x1=2; and so, in one rotation in the tertiary plane, A is first at **A**, then at **B**, and that's the end of one rotation. In drawing #34, we see that formula illustrated in the case of a form native to, and currently at, OD2. 2x2=4; and so, in one rotation in the tertiary plane, A is first at **A**, then at **A'**, then at **B**, then at **C**, and that's the end of one rotation in the tertiary plane. Surely my readers can depict for themselves how rotations in the tertiary plane shall work at OD3,

ESOPTRICS: LOGIC OF THE MIRROR

OD4, etc.! Confident of that, I'll move on to the primary and secondary planes.

Col #1	Col #2	Col #3	Col #4	Column #5	Column #6
D_{nc}	Plane 3	Plane 2	Plane 1	$S_1 \times S_2$	=
	$S_3 = 2D_{nc}$	$S_2 = \sqrt{D_n}\updownarrow$	$S_1 = \sqrt{D_n}\updownarrow$	S Plane 1 per Plane 3	
1	2	1	1	1 @ 1	= 1
2	4	1	2	1 @ 2	= 2
3	6	2	1 or 2	1 @ 2 + 1 @ 1	= 3
4	8	2	2	2 @ 2	= 4
5	10	2	2 or 3	1 @ 2 + 1 @ 3	= 5
6	12	2	3	2 @ 3	= 6
7	14	3	2 or 3	2 @ 2 + 1 @ 3	= 7
8	16	3	2 or 3	1 @ 2 + 2 @ 3	= 8
9	18	3	3	3 @ 3	= 9
10	20	3	3 or 4	2 @ 3 + 1 @ 4	= 10
11	22	3	3 or 4	1 @ 3 + 2 @ 4	= 11
12	24	3	4	3 @ 4	= 12
13	26	4	3 or 4	3 @ 3 + 1 @ 4	= 13
14	28	4	3 or 4	2 @ 3 + 2 @ 4	= 14
15	30	4	3 or 4	1 @ 3 + 3 @ 4	= 15
16	32	4	4	4 @ 4	= 16
17	34	4	4 or 5	3 @ 4 + 1 @ 5	= 17
18	36	4	4 or 5	2 @ 4 + 2 @ 5	= 18
19	38	4	4 or 5	1 @ 4 + 3 @ 5	= 19
20	40	4	5	4 @ 5	= 20
21	42	5	4 or 5	4 @ 4 + 1 @ 5	= 21
22	44	5	4 or 5	3 @ 4 + 2 @ 5	= 22
23	46	5	4 or 5	2 @ 4 + 3 @ 5	= 23
24	48	5	4 or 5	1 @ 4 + 4 @ 5	= 24
25	50	5	5	5 @ 5	= 25
26	52	5	5 or 6	4 @ 5 + 1 @ 6	= 26
27	54	5	5 or 6	3 @ 5 + 2 @ 6	= 27
28	56	5	5 or 6	2 @ 5 + 3 @ 6	= 28
28	58	5	5 or 6	1 @ 5 + 4 @ 6	= 29
30	60	5	6	6 @ 5	= 30
31	62	6	5 or 6	5 @ 5 + 1 @ 6	= 31
32	64	6	5 or 6	4 @ 5 + 2 @ 6	= 32

CHART #5: TRIPARTITE STRUCTURE OF PATTERNS OF DIVINE ROTATION.

Column #1 gives us the native and current OD of the form to which the tripartite pattern belongs. Column #2 gives us the number of states, positions, locations, places, and sequential shifts which the given pattern of divine rotation shall

150

produce in every complete rotation in the ***tertiary*** plane *at the given OD*. Column #3 gives us the number of states, positions, locations, places, and sequential shifts which the given pattern of divine rotation shall—regardless of the *current* OD—*always* produce in every complete revolution in the ***secondary*** plane, which is the same as saying: in the course of each *segment* in the *tertiary* plane. Column #4 gives us the number of states, positions, locations, places, and sequential shifts which the given pattern of divine rotation shall—regardless of the *current* OD—*always* produce in every complete rotation in the ***primary*** plane, which is the same as saying: in the course of each *segment* in the *secondary* plane. Column #5 shows us precisely how—in the course of one complete *rotation* in the *secondary* plane (which is the same as saying: "in the course of each *segment* of the *tertiary* plane")—the given pattern of divine rotation produces—in the *primary* plane—a total number of states, positions, locations, places, and sequential shifts equaling the native OD of the form to which the pattern of divine rotation applies. For example, line 1 of column 5 tells us that, for each state in the tertiary plane, there is, in the primary plane, always 1 rotation consisting of only 1 state (***i.e.:*** in effect, no rotation at all); line 2 of column 5 tells us that, for each state in the tertiary plane, there is no rotation in the secondary plane and, in the primary plane, one rotation consisting of 2 states.

The most important bit of information given to us by chart #5 is this: If you're trying to tell whether you're dealing with a rotation (***i.e.:*** positive rotation) or with a reverse one (***i.e.:*** a negative rotation), you cannot do so in the *primary* plane, until you reach the form which is native to OD6, because only there is every rotation in the primary plane definitely either positive or negative. In the *secondary* plane, you cannot make that distinction until you're dealing with the form which is native to OD9, because only there is every rotation in both the primary and secondary planes definitely either positive or negative. Above all, what that means is this: The rotations of the three *categorical* forms native to OD1, OD2, and OD4 are *never* either positive or negative in either the *primary* or the *secondary* planes, and, as long as it remains at OD1, the rotations of the *categorical* form native to OD1 are neither positive nor negative even in the *tertiary* plane. To me, the latter quickly implies that, when forms native to OD1 accelerate beyond OD1, they all, without exception, rotate in the same one manner in the tertiary plane. That same one manner of rotation may be what I call a positive one, or it may be what I call a negative one. I do not pretend to know which way it is.

What's so important about this distinction between forms which, in the primary and secondary planes, have a definite kind of rotation and those which do not? It shall allow us to forge a distinction which shall almost exactly match modern Science's distinction between leptons and quarks. For, on the one hand, we now have 3 categorical forms (***i.e.:*** OD1, OD2, & OD4) which—when accelerated to become the leading form of a kind of sub-atomic super matrix—shall resemble the 3 leptons called electrons, muons, and taus and, on the other, we have 6 categorical forms (***i.e.:*** OD16, $OD2^8$, $OD2^{16}$, $OD2^{32}$, $OD2^{64}$, and $OD2^{128}$) which—when accelerated to become the leading form of a kind of sub-atomic super matrix—shall resemble the 6 quarks. More on this later!

B: THREE KINDS OF ALEPHONS & ANTI-ALEPHONS:

It's been repeatedly said that the ultimate constituents of the universe are each a single carrying generator logically joined at the center of its primary act to the center of the primary act of a single piggyback form. Let's now give a name to each instance of this duplex structure. Hereafter, every instance of a single form concentric with the act of its carrying generator shall be called an "alephon". A multitude of alephons concentric with one another is a "super matrix".

We've repeatedly said there are 2^{256} different kinds of alephons distinguished from one another by their native OD. In studying the tripartite structure of the patterns of divine rotation, we now see how each alephon's *native* OD is indicated by the way its *formal* side (*i.e.:* the piggyback form as opposed to its carrying generator) rotates in its primary and secondary planes of rotation.

In that study, we've also seen that, in most cases, rotations in the primary, secondary, and tertiary planes can be either positive or negative. Immediately, that forces us to conclude that some alephons are positive and some negative, which is to say some are alephons and others anti-alephons.[1] Are they interchangeable? Under some circumstances, can alephons become anti-alephons and vice versa? I doubt it; but, if so, I dare suspect that, wherever some alephon becomes an anti-alephon, then some other anti-alephon must become an alephon. Why?! I suspect the number of positive alephons in the universe versus the number of anti-alephons must always remain constant.[2] If that's the case, then I, for one, can think of only one way to maintain that balance—namely: There must be what some scientists reject as "spooky action at a distance". As I've tried to explain, though, there's nothing spooky about action at a distance in a universe which is only *logically* rather than *physically* extended.

Doubled so far, then, is the number of ultimate constituents native to each of the ontological distances from OD2 thru and including OD2^{255}-1, because half of them are positive (*i.e.:* alephons), and the other half are negative (*i.e.:* anti-alephons). We shall now triple that, and say there are 6 kinds of ultimate constituents (*i.e.:* 3 kinds of alephons & 3 kinds of anti-alephons) native at least to the 6 categorical ultimate constituents native to OD16, OD2^8, OD2^{16}, OD2^{32}, OD2^{64}, OD2^{128}. How shall that happen?

We've repeatedly said that the tertiary plane involves rotations and anti-rotations around the axis B'C'. Since there are 2 other axes, why should rotations in the tertiary plane always be around the axis B'C'? Why can they not sometimes be

---

[1] Since at least 1961, I have maintained that Esoptrics' alephons and anti-alephons are what modern Science calls neutrinos and anti-neutrinos. For Esoptrics, that necessarily means that every occupant of the universe is ultimately a neutrino or an anti-neutrino or some kind of combination of neutrinos and/or anti-neutrinos. Since what's *without* mass can't add up to something *with* mass, that demands one conclude that neutrinos and anti-neutrinos have mass, and that's what I maintained for decades in the face of modern Science's assertion that neutrinos and anti-neutrinos have no mass. It took quite a while; but, modern Science has finally caught up with me and now admits that neutrinos and anti-neutrinos have mass.

[2] There may be a slight imbalance due to the fact that there is only one kind of alephon native to OD1 and OD2^{256}. As for OD1 alephons, are they all alephons, or are they all anti-alephons? I do not know. All I do know is: Either they are all alephons, or they are all anti-alephons.

around the axis AB and sometimes around the axis A'C? They can, but not for the same alephon or same anti-alephon; and so, there must be 3 kinds of alephons and 3 kinds of anti-alephons, and the difference between one kind and another is the axis of rotation in the tertiary plane. If, for some given alephon or anti-alephon, its rotations and anti-rotations in the tertiary plane are around the B'C' axis, then that's the way it *forever* is for that given alephon or anti-alephon. The same holds true wherever the axis is either AB or A'C.

In my book ***Introspective Cosmology II***, I included the following chart:

AXIS OF ROTATION IN:

SPECIES:	TERTIARY PLANE:	SECONDARY PLANE:	PRIMARY PLANE:
Aleph$_1$	AB	A'C	B'C'
Aleph$_2$	AB	B'C'	A'C
Beth$_1$	A'C	AB	B'C'
Beth$_2$	A'C	B'C'	AB
Daleth$_1$	B'C'	A'C	AB
Daleth$_2$	B'C'	AB	A'C

CHART #6: SIX WAYS TO ARRANGE THE THREE PLANES.

I now reject the above chart as unduly complicated, since I cannot now see how it makes any difference what the axis of rotation is in either the secondary or the primary planes. For example, what difference does it make if Aleph's axis of rotation in the second plane is B'C' rather than A'C? As long as the axis of rotation in the tertiary plane is AB, it now seems to me that it's irrelevant how the remaining 2 axes function. These days, then, I am concerned only with the 3 ways to arrange the *tertiary* plane. For Aleph alephons and anti-alephons, the tertiary axis of rotation is AB; for Beth Alephons and anti-alephons, the tertiary axis of rotation is A'C; and for Daleth Alephons and anti-alephons, the tertiary axis of rotation is B'C'.

But, what's the point in introducing these 3 ways to rotate or anti-rotate in the tertiary plane? Let's go back to drawing #27 on page 137. It's Geometry's way of depicting the quest for neutrality at the most basic level. Now that you've learned something about rotations and anti-rotations in the three planes, you should be able to imagine quite readily how neutrality applies to drawing #27. First, +A must represent a particle (*i.e.:* a super matrix composed of many concentric alephons) rotating around its own B'C' axis, and –A must represent +A's anti-particle. As +A's anti-particle and mirrored image, -A should be *anti*-rotating around its own B'C' axis. That, though, is not going to produce neutrality at the nexus, ⊛, unless: (1) +A, while rotating around it own B'C' axis, is also rotating around the larger square's B'C' axis, and (2) -A, while anti-rotating around it own B'C' axis, is also *anti*-rotating around the larger square's B'C' axis. But, how can they do that if the larger square is a single particle? Obviously, they cannot.

It must be, then, that the larger square represents 2 particles, one of which is rotating around their common B'C' axis while the other is anti-rotating around that same axis. But, that requires this: In each of these 2 super matrices, the leading form is at some same, one current OD. Can 2 forms and the carrying generator of each be simultaneously at the same OD???!!! For decades I wrestled with that question. Only recently did I finally say to myself: "By virtue of the way they rotate around their common B'C' axis (*i.e.:* A→A'→B→C vs. A→C→B→A'), these two leading forms are more than adequately distinct from one another *logically*; and so, they can indeed reside simultaneously at the same OD."

That solves that issue. Immediately, though, another arises: With +A and –A spinning oppositely and orbiting the nexus, ☻, in opposite directions, that produces neutrality around only one of the three axes found in every form. In other words, in what we've just said, we've produced a tripartite particle which is only one-third neutral and two-thirds non-neutral. Therefore, to achieve *total* neutrality (*i.e.:* three-thirds neutrality and zero non-neutrality), the above scenario must be repeated around the two remaining axes—namely: AB and A'C. How can that be done, if B'C' is the tertiary plane's axis of rotation for *all* forms? By logical necessity, it cannot; and so, by logical necessity, there must be three kinds of forms distinguished from one another by what constitutes their axis of rotation in the tertiary plane. In turn, each of these 3 kinds must admit of 2 sub-kinds distinguished from one another by the way they spin around their common axis. Thus do we arrive at:

NAME	TERTIARY PLANE'S AXIS OF ROTATION	SPIN METHOD
+Aleph	AB	A'→B'→C→C'
-Aleph	AB	A'→C'→C→B'
+Beth	A'C	A→B'→B→C'
-Beth	A'C	A→C'→B→B'
+Daleth	B'C'	A→A'→B→C
-Daleth	B'C'	A→C→B→A'

CHART #7: THE 3 POSITIVE & 3 NEGATIVE SUB-SPECIES OF ALEPHONS.

Possibly, I do not have to mention that I have no way to be sure I've got the spin methods properly assigned. It may well be that the method I've assigned to +Aleph should be assigned to –Aleph and vice versa. By the same token, +Beth and –Beth may be backwards. The same can be said of +Daleth and –Daleth. Someone else shall have to decide which spin method is most properly called negative and which most properly called positive.

Think about it for a moment, and there can be no such thing as a super matrix with a leading form composed of only 1 positive and 1 negative alephon of the same sub-species. For example, there can be no leading form composed only of +Aleph and –Aleph. Why?! In such a case, the leading form would be leading a super matrix which is one-third neutral and two-thirds non-neutral; but, there would be no polarity, because the only rotation is in the tertiary plane and is in both directions. Conversely, suppose the leading form is solely a +Aleph. In that case, it's a *positive* three-thirds non-neutral. Again, suppose the leading form is composed of a +Aleph, a –Aleph, and a –Daleth. In that case, it's a *negative* two-thirds non-neutral, which is to say a *negative* one-third neutral. If it's composed of a +Aleph, a –Aleph, a +Beth, a –Beth, and a +Daleth, then it's a *positive* one-third non-neutral and a *positive* two-thirds neutral. Finally, if it's composed of all 6 sub-species, then it's a neutron, which is to say a particle which is three-thirds neutral and without either a positive or a negative polarity.

At what ontological distances can these 6 sub-species be found? When it comes to ontological distances above 2^{129}, my mind is made up to no extent whatsoever. My mind is fairly well made up with regard to OD's below 2^{129}. That brings us to the next section.

C. MONONS, DIONS, & SEXTONS:[1]

Every alephon native to OD1 is a monon. I use the term "monon" because it applies to only one categorical OD. When one accelerates to become the leading form in a sub-atomic super matrix (one which is necessarily composed solely of monons), the resulting particle is also a monon. Science calls it an electron.

Every alephon native to either OD2 or OD4 is a dion. I use the term "dion", because it applies to only two categorical OD's. If a form native to OD2 accelerates to become the leading form in a sub-atomic super matrix (probably it's composed solely of dions), the resulting particle is also a dion. If a form native to OD4 accelerates to become the leading form in a sub-atomic super matrix (probably it's composed solely of dions), the resulting particle is also a dion.

Every alephon native to either OD16 or $OD2^8$ or $OD2^{16}$ or $OD2^{32}$ or $OD2^{64}$ or $OD1^{128}$ is a sexton. I use the term "sexton" because it applies to 6 categorical OD's. Wherever one of these is the leading form in a sub-atomic super matrix, the resulting particle is a sexton.

Monons and dions can be looked upon as 2 species within a genus called "sub-sextons". That's because all monons and dions are necessarily three-thirds non-neutral. Why is that?

Remember what we saw in chart #5 on page 150: For alephons native to OD1, OD2, OD3, and OD4 there are either no rotations at all or no *definitive* rotations in either the primary or the secondary planes, and, for alephons native to OD1, not even the tertiary plane has a *definitive* rotation at OD1. That first means: (1) All alephons native to OD1 rotate the same way in the tertiary plane no matter what their current OD; (2) half the alephons native to OD2 or OD3 or OD4 rotate and the other half anti-rotate; but (3) whether at OD2 or OD3 or OD4, there's no such thing as both an alephon and an anti-alephon joined together at the same OD. Dions either rotate in the tertiary plane or they anti-rotate, but not both. Otherwise, there's no polarity. That, in turn, necessarily means this: Dions, like monons, cannot achieve neutrality with regard to even so little as 1 of their 3 axes; and so, all

---

[1] In my book **Introspective Cosmology II**, I had much to say about what I called "octons". I have since rejected that concept as impossibly complicated.

dions, like all monons, must always be three-thirds non-neutral. If the dion's rotation in the tertiary plane is positive, then it's a positive three-thirds non-neutral. If the dion's rotation in the tertiary plane is negative, then it's a negative three-thirds non-neutral.

Because monons always have no rotations *whatsoever* in their primary and secondary planes, they preclude the existence of Aleph and Beth alephons native to OD1. Because they have no *definitive* rotations in the tertiary plane in their native state, there can be no such thing as both +Daleth and –Daleth alephons native to OD1. Either all monons are +Daleth monons, or all monons are –Daleth monons. For now, I do not know how to decide which alternative rules.

Because, in their primary and secondary planes, dions have either no rotations whatsoever or no *definitive* ones, they preclude the existence of Aleph and Beth alephons native to OD2, 3, or 4. As we said, because they have a definitive rotation in the tertiary plane, half the alephons native to those OD's can be +Daleth and the other half –Daleth alephons. Nevertheless, because all dions have only one axis of rotation, the need for polarity rules out the possibility of a dion in which both a +Daleth and a –Daleth alephon are present at the same OD. Half the dions, then, are positive and the other half negative, and never the twain shall meet.

The sub-sextons are either monons or dions. All the sextons are sextons; and so, they are *sui generis*, as they say. As we've seen, they are native to one of 6 categorical OD's, and there are 3 positive and 3 negative ones at each of those 6 OD's. That gives us a total of 18 positive and 18 negative sextons, and the difference between the positives and negatives is the nature of their curvilinear motion.

In sum, then, going by chart #5 on page 150, only 1 of the 3 positive and 1 of the 3 negative sub-species of alephons can ever be found *below* OD9 and never in combination at the same OD. All 3 of the 3 positive and all 3 of the 3 negative sub-species are always found *above* OD9. It remains to be seen, though, exactly where and to what extent they are to be found in combo at the same OD. As for whether or not, or to what extent, they are found beyond $OD2^{129}$-1, I do not currently know. Still, whenever I hear of heavenly bodies orbiting a star in a manner the reverse of the way Earth orbits the Sun, I can't help but suspect that at least one of the alephons and its anti-alephon is native to every OD beyond $OD2^{129}$.

D: EARTH AS AN EXAMPLE OF
HOW THE PATTERNS OF DIVINE ROTATION CAUSE FORMS TO PRODUCE
CURVILINEAR MOTION AT THE CELESTIAL LEVEL:

Let me now try to give my readers a specific example of how the patterns of divine rotation produce curvilinear motion on a celestial scale. To that end, let's talk about how Earth rotates around its polar axis and orbits the Sun. Figure #35 on the next page is an attempt to help my readers get a clear grasp of what is about to follow.

As we start, be sure to bear in mind two urgings which have been repeated *ad nauseam*: (1) One must imagine the square is a cube with B' in the air above the sheet and C' in the air underneath the page; and (2) this, too, is a case of using Geometry's pictures to convey Algebra's abstract principles.

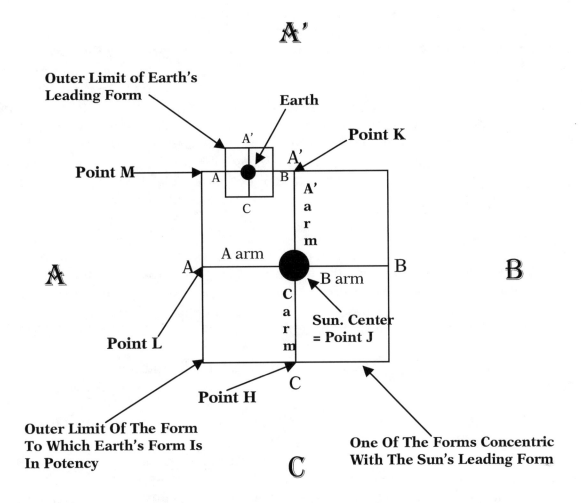

Drawing #35: How Forms Produce Earth's Curvilinear Motions.

According to the 1991 edition of the ***Encyclopedia Americana***, on page 192 of volume 25, as Earth orbits the Sun, its greatest distance is 94,510,000 miles, and its least distance is 91,400,000 miles. The mean distance is given as 93,000,000 miles.

On page 657 of volume 29 of the 1998 edition of the ***Encyclopedia Americana***, we are told that the tropical year (also called the solar year) is the length of time it takes for Earth to go from one vernal equinox to the next, and this length of time is calculated to be 31,556,925.9747 seconds. At least, that's the way it was for the tropical year for Dec. 31, 1899. We are told a specific date had to be used because the tropical year is decreasing in length by less than 1 second per century. I will now try to show my readers how those figures apply to the above drawing.

At the center of Earth is a cosmic super matrix (***i.e.:*** black hole?) held together by its leading form. I do not pretend to know either what that leading form's current OD is or what its native OD is. Lack of such knowledge shall to no extent detract from my ability to achieve my current goal. For now, all we need know is that the small square represents that leading form regardless of what its native or current OD's are.

At the center of the Sun is another cosmic super matrix held together by its leading form. Here again, I do not pretend to know either what that leading form's current OD is or what its native OD is. Here again, that's knowledge we don't presently need. For now, all we need know is that the larger square *does not* represent that leading form; rather, it represents one of the forms made concentric with that leading form. More importantly, it represents the form to which Earth's leading form is in potency. If you remember, that means this: Thru its carrying generator, Earth's leading form is using that solar form as what Geometry speak would call: "the space envelope providing Earth's carrying generator with a field of locomotion". Esoptrics would call it: "a logical cube of potential acts".[1]

Focusing on the *smaller* square, ask this: What will happen if this form's 4 heavenly points A, A', B, and C play "ring around the Rosie" around the B'C' axis? Answer: No matter where point A is, the A arm of the form must be aligned with it; no matter where point A' is, the A' arm of the form must be aligned with it; no matter where point B is, the B arm of the form must be aligned with it; and no matter where point C is, the C arm of the form must be aligned with it. I dare say you can easily perceive the implication: As A, A', B, and C play ring around the B'C' Rosie, the smaller square rotates with them and, thereby, causes Earth to rotate around its polar axis.

Focusing on the *larger* square, ask this: What will happen if this form's 4 heavenly points A, A', B, and C play "ring around the Rosie" around the B'C' axis? Again, the answer is: No matter where point A is, the A arm of the form must be aligned with it; no matter where point A' is, the A' arm of the form must be aligned with it; no matter where point B is, the B arm of the form must be aligned with it; and no matter where point C is, the C arm of the form must be aligned with it. Again, I dare say you can easily perceive the implication: As A, A', B, and C play ring around the B'C' Rosie, the larger square rotates with them and, thereby, carries the Earth around the Sun. That's because, throughout the larger square's every rotation around its center in the Sun, Earth stays rather close to where it is relative to the A' arm and point K.

Just how close does Earth remain to point K? To answer, we must exactly determine the native and current OD's of the form to which Earth's leading form is in potency. For Esoptrics, then, the larger square represents a form whose *native* OD is exactly 2^{143} (*i.e.:* 1.1150372599 x 10^{43}). Its *current* OD is very close to being 2^{197} (*i.e.:* 2.0086725553 x 10^{59}). How close?! Its *current* OD is at 97.81957103313% of 2^{197}. For the moment, we shall round that off at 97.82% and say the current OD of the form is: 1.9648834936 x 10^{59}.

---

[1] Because Earth's every orbit of the Sun is the same, Esoptrics says the form to which Earth's leading form is in potency is concentric with the Sun's leading form. When, in the case of Mercury, a planet's orbits are *not* all the same, Esoptrics says the form to which the planet's leading form is in potency is *not* concentric with the Sun's leading form; instead, it's one moving around near the Sun's center. Naturally, as it moves around, the center around which the planet rotates moves and, thereby, causes the planet to have several different orbits. I have not the skill to calculate *exactly* how that works.

According to formula #1 on page 36, if a form is exactly at $OD2^{197}$, then its LB&D = $2D_c$, where D_c = current OD. Double 2^{197}, and you have $2^{198}\Phi$, which is to say: $4.01734511064 \times 10^{59}\Phi$. On page 31, 1 cm. = $1.361129468 \times 10^{46}\Phi$. Divide the first number by the second, and: $(4.01734511064 \times 10^{59}) \div (1.361129468 \times 10^{46})$ = 29, 514,790,500,000 cm.. Divide that by 160,934.72 (*i.e.:* cm. per mi.), and we learn that the LB&D of a form exactly at $OD2^{197}$ = 183,396,040.95374 mi.. But, the form is not *exactly* at $OD2^{197}$; therefore, we multiply 183,396,040.95374 by 97.82%, and we get: 179,398,007.26094 mi. = the LB&D of the form at 97.82% of $OD2^{197}$. For drawing #35 on page 157, that means lines AB and A'C are each 179,398,007.26094 mi., and lines MA' and A'J are each half that length at 89,699,003.63047 miles. Since 89,699, 003.63047 miles is less than 91,400,000 miles (*i.e.:* Earth's nearest approach to the Sun according to ***Encyclopedia Americana***), Earth must always be somewhere to the left or the right of point K in drawing #35. We've made some progress; but, we still don't know how far to the left or right of point K.

Draw a line from the center of the Sun to some point to the left of point K. Call this new point "W". It represents the distance of 91,400,000 miles given by ***Encyclopedia Americana*** as Earth's minimum distance from the Sun. Doing that, you create a right triangle whose sides are WA', A'J, and JW. We already know the lengths of JW and A'J. The former is 91,400,000 mi. and the latter 89,699,003.63 047. Line WA' is the unknown. How shall we calculate its length?

Since we're dealing with a right triangle, line JW is the hypotenuse of a right triangle, and, as the Pythagorean theorem tells us: The square of the hypotenuse = the sum of the squares of the other 2 sides; therefore: $(JW)^2 = (WA')^2 + (A'J)^2$, which is to say: $(91,400,000)^2 = X^2 + (89,699,003.63047)^2$. That works out to: $X^2 = 8,353,$ 960,000,000,000 minus 8,045,911,248,626,793. X^2 then = 308,048,751,373.207 and X = the square root of that latter number, and that square root = 17,551,317.653 mi.. In short, point W is 17,551,317.563 miles to the left of point K.

Draw a second line from the center of the Sun to a point left of both point K and point W. Call it point Y, and the length of this new line, JY, is 93,000,000 mi., which is to say the mean distance of Earth from the Sun. Do the Pythagorean theorem bit again, and the length of line KY is 24,557,865.3667864 miles. Since the length of line MA' has already been established at 89,699,003.6 miles, we're still not even half way from point K to point M.

Draw a third line from the center of the Sun to a point left of both point K, point W, and point Y. Call it point Z. The length of this new line, JZ, is 94,510,000 mi., which is to say the maximum distance of Earth from the Sun. Do the Pythagorean theorem bit again, and the length of line KZ is 29,769,596.09019 miles. We're still not half way from point K to point M.

For Esoptrics, Earth moves toward and away from the Sun mainly as a result of shuttling back and forth from point W to point Z to point W. In one half of its orbit, it's moving from point W to point Z, and, in the other half, it's moving from point Z to point W. Why does it reverse course? The cause is the pulsing motion of the form to which it is in potency. Yes, this is a throwback to the ancient theory that, as the planets orbit their common center, each performs an epicycle around a center unique to itself. As I said earlier, it is also a throwback, in some sense, to the ancient theory that the heavenly bodies perform their curvilinear motions under the influence of angels. O Science, if Esoptrics is right, how great a quantity of egg you shall be found wearing on your face!

Thus far, we've seen how Esoptrics agrees with Science exactly on the size of Earth's orbit. What about how long it takes for Earth to make one complete rotation around its center in the Sun? That amounts to asking this: If a form is native to $OD2^{143}$ and is currently at 97.81957103313% of $OD2^{197}$, how long does it take to complete one rotation in the tertiary plane? To keep the calculations as simple as possible, let's assume, for a moment, that the form native to $OD2^{143}$ is exactly at $OD2^{197}$.

On page 149, formula #7 (*i.e.:* $S_3 = 2D_c$) tells us that, for a form currently at $OD2^{197}$, the number of states, shifts, positions, etc., in the tertiary plane equals 2^{198}, which is to say $4.01734511 \times 10^{59}$. As chart #5 on page 150 illustrates, it's been repeatedly said that, no matter what its *current* OD, every form's number of states, shifts, etc., in the primary plane = D_n per shift in the tertiary plane, where D_n = the form's native OD. Therefore, if a form currently at $OD2^{197}$ is native to $OD2^{143}$, then, in the course of one complete rotation in the tertiary plane, the total number of shifts in the primary plane = $2^{143} \times 2(2^{197})$ = $(1.1150372599 \times 10^{43}) \times (4.01734511 \times 10^{59})$ = $4.4794894835 \times 10^{102}$ shifts. Since shifts in the primary plane are at the rate of one per alphakronon, the duration of each complete rotation on the part of a form native to $OD2^{143}$ and currently at $OD2^{197}$ = $4.4794894835 \times 10^{102}$K.

But, our form is *not exactly* at $OD2^{197}$; rather, it's at 97.81957103313% of it. So, multiply ($4.4794894835 \times 10^{102}$) by .9781957103313. Do that, and the result tells us the duration of each rotation of our form is $4.3818173972338 \times 10^{102}$K. Let's now convert that last figure to seconds. On page 31, one sec. = $1.388543802 \times 10^{95}$K. Therefore, to obtain the duration in seconds of our form's every complete rotation in its tertiary plane, we calculate: ($4.3818179372338 \times 10^{102}$) ÷ ($1.388543802 \times 10^{95}$) = 31,556,925.974696 seconds.

Remember what we said on page 157: According to the ***Encyclopedia Americana***, the duration of one tropical (*i.e.:* solar) year is 31,556,925.9747. Here again, Esoptrics is *exactly* in line with what Science claims to have observed.

Some may object: "From what Esoptrics says, Earth's orbit around the Sun must be a very jerky one." Let's try to answer that question in mathematical terms which are fairly precise. To ease the burden of that task, I'm going to proceed using 4 slightly erroneous assumptions: (1) Currently, the form native to $OD2^{143}$ is *exactly* at 2^{197};[1] (2) the length of lines KJ and MA' is 83,181,864 miles; (3) Earth, at 93,000,000 miles from the Sun is half way between points M and A'; and (4) the length of an Earth year is 31,556,880 seconds, which is to say 365 days, 5 hours, and 48 minutes. Between these calculations and the previous ones, the difference is so small, it shall have only a negligible effect upon the lessons about to follow. In what's about to follow, we're talking about rotations on the part of the larger square.

Each time A, A', B, and C make one complete rotation in the tertiary plane, each will occupy $2 \times 2^{197} = 2^{198}$ (*i.e.:* approximately 4.01734×10^{59}) distinct absolute

---

[1] In capturing this form, the Sun's leading form raised this form's OD some 2^{54} times from its native one of 2^{143} to its current one of 2^{197}. As a captured form, it does not itself capture other forms as it accelerates from $OD2^{143}$ to 2^{197}.

frames of reference. If one complete rotation of the Earth around the sun covers 584,000,000 miles (*i.e.:* 93,000,000 × 2π), then multiplying that by 1.609347 gives us a figure of 939,858,640 km. which translates into 939,858,640,000 meters or 93,985, 864,000,000 cm.. Divide that number of centimeters by the number of absolute frames of reference, and you have: 93,985,864,000,000 cm. ÷ (4.01734 x 10^{59}) = 2.3395 x 10^{-45} cm.. In turn, that tells us that, whenever each of the points A, A', B, and C instantaneously shift from one absolute frame of reference to another, they instantaneously shift a distance of roughly 10^{-45} cm. (So Geometry speak would describe it.). I dare say no one on this planet yet has the technology to detect a jump that small; consequently, we should not worry ourselves about the "jerkiness" of the curved motions produced by the patterns of divine rotation.

Now see what the duration of each earthly year tells us: If the Earth makes a complete rotation around the Sun in 365 days, 5 hours, and 48 minutes, then that translates into 31,556,880 seconds. Applied to what we just said in the prior paragraph, that would mean that, in figure #35, each of the 4 heavenly points, A, A', B, and C would make their 4.01734 x 10^{59} consecutive leaps (*i.e.:* the number in one rotation in the tertiary plane) in the course of 31,556,880 seconds. How many times, then, does each of them leap per second? Divide (401,734,000 x 10^{51}) by 31,556,880, and we discover each of them leaps 1.273 x 10^{52} times per second. From that it follows that each of those four heavenly points jumps 10^{-45} cm. approximately once every 10^{-52} seconds (So Geometry speak would describe it.). Here again, we are talking about something our current level of technology offers us no possibility of detecting. As I said: We should not worry ourselves about the "jerkiness" of the curved motions produced by the patterns of divine rotation.

We've seen, then, how the larger square rotates through the tertiary plane. However, rotation in the tertiary plane is but an astronomically insignificant part of all that's involved in a *complete* pattern of divine rotation. We have yet to speak of rotation in the secondary and primary planes. Let's now move to the secondary plane.

Focus again on the large square, and imagine that A, A', B, and C have each swung 10^{-45} cm. in a clockwise direction around B'C'. Needless to say, that's a change so slight, we cannot even begin to reflect it in our illustration. Still, each _is_ in a new position which it will occupy for only about 10^{-52} seconds. In that astronomically brief time frame, however, an incredibly vast quantity of activity is going to take place.

First of all, A, C', B, and B' will make one complete rotation around the A'C axis. Looking at figure #35 on page 157, you can see that will necessarily cause the Earth to make a complete rotation around the point where the A' arm intersects the outer perimeter of the larger square. The drawing calls that "point K". Since A, C', B, and B' will make one complete rotation around A'C in the 10^{-52} seconds allowed them, Earth will make one complete rotation around point K in 10^{-52} seconds.[1]

---

[1] At the risk of confusing my reader, let me toss in one, very important point: As the larger square rotates around its own A'C axis, it does not thereby cause Earth's leading form to rotate around _its_ own A'C axis. Earth's leading form rotates around its own 3 axes only to the extent dictated by the heavenly sextet to which it and it alone is logically connected. By way of example, when a Ferris

How great a distance will it cover in that 10^{-52} seconds? The answer, of course, is fairly close to: 41,590,864 miles × 2π = 261,323,105 miles × 1.609347 = 420,559,555 kilometers = 42,055,955,500,000 centimeters. Are you wondering how I arrived at that figure?

In figure #35, the distance from point K to the center of Earth equals one-half the length of any one of the four arms. Draw a line from the center of the Sun (*i.e.:* the center of the larger square) to the center of the Earth (*i.e.:* the center of the smaller square), and you will produce a right triangle the hypotenuse of which is the last line you just drew. If science is correct in telling us the mean distance of the Earth from the Sun is approximately 93,000,000 miles, then the length of our right triangle's hypotenuse is 93,000,000 miles, and the length of the other two sides is X (*i.e.:* the distance from point K to the center of Earth) and 2X (*i.e.:* the distance from point K to the center of the Sun). Since the square of the hypotenuse equals the sum of the squares of the other two sides, we know this: 8,649,000,000, 000,000 (*i.e.:* $[93,000,000]^2$) = $5X^2$ (*i.e.:* $X^2 + 4X^2$). Dividing the square of the hypotenuse by 5 gives us the value of X^2 as 1,729,800,000,000,000 miles, and from that, it follows that X (*i.e.:* the distance from point K to the center of the Earth) = 41,590,864 miles and 2X (*i.e.:* the distance from point K to the center of the Sun) = 83,181,728 miles. Since 41,590,864 is the radius of Earth's rotation around point K, we multiply it by 2π to obtain the circumference of Earth's orbit around point K and that give us: 261,323,105 miles × 1.609347 = 420,559,555 kilometers = 42,055, 955,500,000 cm.. Use a shorthand notation, and the latter is: 4.2×10^{13} cm..

So then, in figure #35, Earth—*in the secondary plane!*—whirls around point K at a rate of 4.2×10^{13} cm. every 10^{-52} seconds. That translates into 4.2×10^{65} cm. per second. Since the speed of light is generally given as 3×10^{10} cm. per second, figure #35 gives Earth a speed—*in the secondary plane!*—of approximately 1.4×10^{55} times the speed of light. That will strike the scientists as ludicrous; but, one must bear in mind that we're using *our* plane's ruler to measure what's happening in *another* plane.

Let us now take an even closer look at Earth's motion around point K. As we did with Earth's motion around the Sun, let us describe the various segments of its motion around point K. To do that, we must examine how the larger square moves through the secondary plane.

We said that the larger square represents a form whose *native* ontological distance is 2^{143} (*i.e.:* approximately 1.115037×10^{43}). What does that tell us about the number of discrete states (*i.e.:* locations, places, positions, shifts, etc.) in the secondary plane of that form. On page 150 and near the top of column 3 in chart #5, we were given what we shall now call:

Formula #8: $S_2 = \sqrt{D_n}\updownarrow$

Since S_2 = number of states in the secondary plane; and D_n = the native OD of the form in question; and $\updownarrow$ = rounded up or down to the nearest whole num-

---

wheel rotates $360°$, it carries the attached bucket seats with it, but does not cause the bucket seats to mimic its own rotation; otherwise, the passengers would wind up upside down.

ber—formula #8 tells us that the number of discrete positions in the secondary plane of that form = $\sqrt{2}^{143}$ rounded up or down to the nearest whole number. Since 2^{143} is roughly 1.115037×10^{43}, invoking that formula will give us a result of 3.3392 17363285 $\times 10^{21}$. That tells us that, as Earth swings around point K, it covers a distance of 4.2×10^{13} cm. by means of a total of $3.339217363285 \times 10^{21}$ shifts. Necessarily, then, each shift would involve a change of approximately 1.26×10^{-8} centimeters, and, since 3.339217×10^{21} such shifts take place in 10^{-52} seconds, then the individual shifts would take place roughly every 10^{-73} seconds.

So then, in the *tertiary* plane, the pattern of divine rotation pertinent to the larger square in figure #35 has caused the 4 heavenly points A, A', B, and C to shift instantaneously from their 4 home positions to 4 new absolute frames of reference which can be described as having centers 10^{-45} cm. from the home positions and in a clockwise direction around the B'C' axis. Here, the pattern of divine rotation rests for 10^{-52} seconds in the *tertiary* plane. However, in that 10^{-52} seconds, the pattern of divine rotation switches to the secondary plane and, thereby, causes the 4 heavenly points A, C', B, and B' to make one complete rotation around the A'C axis. It will do such by causing each of the 4 points to rest 10^{-73} seconds in each of 10^{21} successive absolute frames of reference.

Rotation in the *primary* plane now comes into play. In the 10^{-73} seconds allowed in each of the 3.3392×10^{21} locations in the secondary plane, the 4 heavenly points A', B', C, and C' will make one complete rotation around the AB axis. How many states, positions, shifts, etc., shall that involve. The answer to that is what, in this book, shall be called:

Formula #9: $S_1 = \sqrt{D_n}\updownarrow$

I dare say you can readily see the similarity between formula #9 and formula #8 on the prior page. There's no change save that S_2 has become S_1, because we're now dealing with the primary rather than the secondary plane of rotation. Most importantly, there's no significant change in the number of states per rotation. Since the number of positions in each rotation in the *primary* plane equals the same number (or 1 greater or lesser than it) of positions in each rotation in the *secondary* plane, the points A', B', C, and C' will accomplish their rotation by resting 10^{-94} seconds in each of 10^{21} successive absolute frames of reference. What, for Geometry speak, shall be the total distance covered by the complete rotation, and what will be the distance covered by each instantaneous shift?

If you look at figure #35, you should be able to see this: The rotation in the primary plane will cause the planet Earth to rotate around a point on the A arm—a point half way from the center of the Sun to the intersection of the A arm with the outer perimeter of the larger square at point L. Earth shall, of course, rotate around that point at a radius equal to the length of the A' arm. As we have already said, the length of the A' arm equals the length of the A arm, equals the length of the B arm, equals the length of the C arm, and equals 83,181,728 miles. Therefore, the radius of Earth's orbit around the A arm is 83,181,728 miles. Multiplying that by 2π tells us the circumference of that orbit is 522,646,210 miles. To convert that to centimeters we multiply by 160,934.7 and get 84,111,911,000,000 cm. which, in shorthand notation, becomes 8.4×10^{13} cm..

We now know the total distance Earth will travel with each complete rotation in the *primary* plane. As you can readily see, it is *twice* the distance Earth travels with each complete revolution in the *secondary* plane. Of course, each complete rotation in the secondary plane takes a "full" 10^{-52} seconds; whereas, each rotation in the *primary* plane takes "only" 10^{-73} seconds.

What will be the distance covered by each instantaneous shift? Knowing that Earth covers a distance of 8.4×10^{13} cm. by means of a total of 3.3392×10^{21} jumps, we divide the first figure by the latter and determine the distance covered by each jump is 2.52×10^{-8} cm..

If, in the primary plane, Earth travels 8.4×10^{13} cm. in 10^{-73} seconds, then in one second, Earth travels 8.4×10^{86} cm.. Since the speed of light is generally given as 3×10^{10} cm., we know that, in the primary plane, Earth travels at roughly 2.8×10^{76} times the speed of light. Incidentally, 2.8948×10^{76} is 2^{254}.

As we said, one complete rotation in the *secondary* plane will require an "entire" 10^{-52} seconds. In the course of the 10^{-52} seconds required to complete one rotation around point K in the secondary orbit, Earth will circumnavigate the A arm 3.3392×10^{21} times—using $(3.3392 \times 10^{21})^2 = 1.115 \times 10^{43} = 2^{143}$ (*i.e.:* the *native* OD of the form represented by the larger square) successive shifts to do so . In the course of one rotation in the tertiary orbit (*i.e.:* in the course of one Solar year by *our* reckoning) Earth will circumnavigate point K 2^{198} times, which is to say one time at each of the 2^{198} (*i.e.:* roughly 4.01734×10^{59}) locations in the tertiary revolution. At each of those 2^{198} locations, Earth will also circumnavigate the A arm 3.3392×10^{21} times and, therefore, in one solar year, will circumnavigate the A arm $(4.01734 \times 10^{59}) \times (3.3392 \times 10^{21}) = 1.3414 \times 10^{80}$ times—using $(1.3414 \times 10^{80}) \times (3.3392 \times 10^{21}) = 4.4792 \times 10^{101}$ successive shifts to do so. It's nothing more than the kind of simple mathematics which virtually any freshman in high school can employ with reasonable ease.

E: THE STROBE EFFECT AS ANSWER TO AN OBJECTION:

Someone may object: "You're contradicting yourself. On the one hand, you tell us this: The 4 points A, A', B, and C complete one rotation around the B'C' axis by spending roughly 10^{-52} in each of 2^{198} consecutive absolute frames of reference in the tertiary plane. Then, you immediately turn around and tell us this: The points A and B spend no more than 10^{-73} seconds in each of those 2^{198} consecutive frames, because, while they're waiting to jump from one frame in the tertiary plane to the next, they're busy spending 10^{-73} seconds in each of roughly 10^{21} consecutive absolute frames of reference in the *secondary* plane. At about the same time, you give us an even worse story in the case of the two points A' and C. They, it seems, spend no more than 10^{-94} seconds in each of those 2^{198} consecutive frames in the tertiary plane. After all, while they're waiting to jump from one frame in the tertiary plane to the next, they're busy spending 10^{-94} seconds in each of 10^{42} frames in the primary plane. As a result, in any given second by our reckoning, each of the four points A, A', B, and C spends far less time in any one of the points in the tertiary plane than it does someplace else. Why, then, should our human senses lock in on what goes on in the **_less_** frequented locations?"

The answer, of course, is:

THE STROBE EFFECT.

As you possibly know, if, upon a rotating fan blade, you play a light flashing at the appropriate rate, the blades will seem to come to a halt as long as the light flashes at the proper frequency. That comes to pass despite the easily provable fact that each of the now visibly motionless blades spends less time in the visible location than it does in the other locations around the shaft. What's important is that, with our eyes fixed on a certain location, the fan blades be at that location several times in the course of a second. And, isn't that basically the same thing I've told you? In my illustration above, our eyes are fixed on a location to which the Earth returns 10^{52} times per second. Is that not enough times per second?

If sharp, you'll immediately reply: "By no means! Each 10^{-52} seconds, it returns to a location which has shifted 10^{-45} cm. along in the orbit around the Sun." I, though, merely reply: "So does the observer!" Remember! When the fan blade returns to the same spot several times a second, it's not the same spot relative to the Earth's position in its orbit around the Sun; it's merely the same spot relative to where the observer is standing on the Earth's surface.

But, how are we human beings like a flashing light? Well, each of us is a form which changes from primary act to rimary act at regular intervals. Take the rate at which each of these human forms changes from act to act and compare it to the rate at which heavenly points change locations in the primary plane. You will find that the rate of the latter is astronomically greater than the rate of the former. Because of that disparity, our experience of _rhythmical_ changes is limited to those whose rhythm comes within a reasonable proximity of our own. Because only the rhythmical change within the tertiary plane comes close to our own, we experience what goes on there despite the fact that the ultimate constituents of the universe—like the fan blades—spend far less time in the experienced locations than they do in other locations around two other shafts.

F. THE EFFECT OF A FORM'S PULSATION UPON CURVILINEAR MOTION.

Some readers may object: "Wait a minute! What about this business of the forms changing from primary act to primary act? Didn't you say that, in doing so, they first double the LB&D of their first act and then revert to the half of that in the second act of a 2 phase cycle? If that's so, then, in orbiting the Sun, Earth should drop as close as 93,000,000 ÷ 2 = 46,500,000 miles or should swing as far away as 93,000,000 x 2 = 186,000,000 miles, depending upon which phase of the two phase cycle produces the 93,000,000 miles. Clearly, that does not happen; and so, your theory is quite mistaken?"

Remember: When I gave you that principle, we were talking about forms still at their _native_ ontological distances. In figure #35, we're talking about a form whose _current_ OD is 2^{197} but whose _native_ OD is 2^{143}. Now then, at its native OD of 2^{143}, such a form has an LB&D of .0016384 cm. in the course of the first of the two acts constituting its two-phase cycle. In the course of the second of those two acts, that LB&D doubles to .0032768. That's a change, of course, of only .0016384 cm., and that's always going to be the extent of the change for such a form no matter what its *current* OD. Therefore, at its current OD of 2^{197}, the form controlling Earth's orbit around the Sun, has an LB&D of 93,000,000 miles in the first act of its two phase cycle and then, in the second act of that cycle, increases its LB&D by only .0016384 cm. rather than the change of millions of miles you're trying to attribute to the switch from half cycle to half cycle.

To facilitate one's effort to grasp the above, it might help to contrast the above with a very different scenario *suggested* (at least to *myself*, that is) by the idea of native versus current OD. Imagine, then, that the larger square in figure #35 still represents a form centered on the Sun (or some part of the Sun) but which has a *current* ontological distance of 2^{195} (*i.e.*: one-fourth the 2^{197} we previously attributed to it) and a *native* OD of 2^{205} (*i.e.*: $2^{197} \times 2^{8}$). As you can readily see, we are here dealing with a form which has been *de*celerated 2^{10} from its *native* OD (*i.e.*: $2^{205} \div 2^{195} = 2^{10}$).

If, as we said earlier, the length of the A arm is 83,181,718 miles at 2^{197}, then, at 2^{195}, the length of the A arm is: $83,181,718 \div 4 = 20,795,432$ miles. If, at 2^{197}, the length of the A arm is 83,181,718 miles, then, at 2^{205}, the length of the A arm is: $83,181,718 \times 2^{8} = 83,181,718 \times 256 = 21,294,522,368$ miles. From that, 3 conclusions follow: (1) At its native OD and in the <u>first</u> act of its 2 act cycle, it's A arm (as well as it's A', B, B', C, and C' arms) had a length of 21,294,522,368 miles; (2) at its native OD and in the <u>second</u> act of its 2 act cycle, it's A arm's length doubled to $21,294,522,368 \times 2 = 42,589,044,736$ miles; and (3) while currently at OD2^{195}, its 6 arms should each have a length of 20,795,432 mi. in the first half of its 2 act cycle, and then, in the second half of its 2 act cycle, should instantaneously shift to a length of: $20,795,432 + 21,294,522,368 = 21,315,317,800$ miles.

All of this is in accordance with formula #2 on page 36 which told us: $\Phi_2 = 2(D_c + D_n)$, which is to say that, in the 2nd half of its 2 phase cycle, every form's LB&D (*i.e.*: the length of each of its 3 axes) = 2 times the sum of its current and native OD. Naturally, whatever is the length of each of the *three axes*, then the length of each of the *six arms* would be one-half of that.[1]

If a heavenly body's leading form were in potency to the form just described, that heavenly body would have an incredibly elliptical orbit around the Sun. For, at its closest approach to the Sun, it would be only 20,759,432 miles from the Sun, but, at its farthest distance from the Sun, would be 21,315,317,800 miles away. That wild swing would take place because the leading form's carrying generator always moves toward the outermost limit of the form to which its piggyback form makes it in potency; and so: (1) If that limit suddenly expands, the carrying generator must race away <u>from</u> the Sun; but (2) if that limit suddenly contracts, the carrying generator must race <u>toward</u> the Sun.

---

[1] The above is a reversal of what I maintained in my book, **Introspective Cosmology II**. There, I labored under the mistaken notion that, in the second act of its two-act cycle, every form's LB&D *dropped* by the half of its LB&D at its native OD. It never ceases to amaze me how long, in many cases, I can stare into the face of what Esoptrics is trying to tell me and, yet, see it not.

As you can perhaps readily infer, Esoptrics does indeed have a way to explain why the orbits of some bodies around the Sun are almost perfect circles, while the orbits of other bodies around the Sun are very much *other* than "perfect circles". The difference is merely one of being carried around the Sun by a form with a native OD *lower* than its current one versus one with a native OD *higher* than its current one. Hopefully, it's an inference worth mentioning.

G. THE PATTERNS OF DIVINE ROTATION AS THE SOURCE OF MAGNETIC FIELDS:

We've now seen how the patterns of divine rotation work in the second and primary planes of rotation. The way they work in those higher planes is Esoptric's way of explaining what magnetic fields are—something no one has every managed to do until now. How might those patterns in the upper planes explain magnetic fields? To answer that question, we must first talk about how many dimensions there are in the finite world.

So far, we said there are four dimensions—namely: length, breadth, depth, and ontological distance. Actually, if you wish to be as accurate as possible, you should say we've divided our world into *seven* different dimensions. That's because length, depth, and breadth are more accurately described as left, right, forwards, backwards, up, and down. It's the unavoidable result of having six potencies, and that's why, when Esoptrics describes a particular location in space it does so in the terms of all six of the six potencies. To that, ontological distance then adds information regarding the size of what we are describing.

To those *seven* dimensions, the secondary plane of rotation then adds an *eighth* dimension, and the primary plane adds a *ninth* one. Couldn't a numerologist have a ball with that assertion?!

To simplify the relationship between the eighth dimension (*i.e.:* the secondary plane) and the world we experience (*i.e.:* the tertiary plane), one can compare it somewhat to the relationship between the water skier and the waves upon which he rides. Under the guidance of the skier's skillful feet, the water skis do not *collide* with the water; instead, they skim the surface of the liquid mass beneath them. Because the skis do not *collide* with their support, they do not interact with it save—so to speak—after the manner of a feather tickling with the lightest touch imaginable. There is, then, some kind of effect bespeaking the high-speed passage of the water skis; but, the effect is far less impressive and disruptive than that of what happens when the skier "wipes out". In short, there are momentary "tiny tracks in the water" but nothing like the splash which throws gallons of water in every direction and kills perhaps a fish or two, if not the skier himself. Magnetic fields are the "tiny tracks" left in the tertiary plane by the high-speed passage of generators in the secondary plane.

What we have just said of the relationship between our world and the eighth dimension we can also say of the relationship between the eighth dimension (*i.e.:* the secondary plane) and the *ninth* one (*i.e.:* the primary plane). You might say that, in this latter case, one has a water skier skiing atop the waves of water-skiers skiing atop the waves of *our* world.

167

Because the skiers do not *violently* interact with the plane upon which they ski, none of the bodies which move through the secondary plane are bound by the speed restrictions of our seven dimensional tertiary plane, and none of the bodies which move through the primary plane are bound by the speed restrictions of the secondary plane. As has already been mentioned: In the secondary plane, bodies routinely travel at 2^{128} (*i.e.:* 3.4×10^{38}) times the speed of light, and, in the primary plane, routinely travel at the *square* of that, which is to say 2^{256} (*i.e.:* 1.15×10^{77}) times the speed of light. Oh, yes, Captain Kirk! It is quite possible to go from one end of the universe to the other in but a tiny fraction of what *sense experience* tells us is too brief to allow much of anything.

To most scientists, such assertions are laughable nonsense. However, their laughter is made reasonable in their own eyes by a false assumption. Never having engaged in any kind of experience or observation save that which is commonplace in the *tertiary* plane, they falsely assume the parameters used by us to make our observations are the only ones which can ever be used anywhere at any time by any observer whatsoever. I, however, say that some people have minds able to make observations **in** the *secondary* plane. Whenever they utilize that ability, they use a set of parameters so astronomically different from our own, that they, too, come to the conclusion that the maximum velocity within the confines of their *observable* world is the speed of light. By the same token, those who make observations **in** the *primary* plane use yet another set of parameters so astronomically different from those of the secondary plane that, when they apply them, they still come to the conclusion that the maximum velocity within the confines of their *observable* world is the speed of light.

Every intelligent being has a set of parameters which he uses to measure the things which he experiences within his own plane. If you try to use those same parameters to measure or to express the speed of a body moving within another plane, then you will surely arrive at stupefying velocities. All that proves, though, is this: If you're going to talk about velocities in a plane other than your own, then you need to realize you are—as an old adage puts it—comparing "apples and oranges". I can assure my reader that I am well aware of it. It remains to be seen whether or not the scoffers are.

I repeat, then: The distinction between the way the patterns of divine rotation work in the three planes is Esoptric's way of explaining magnetic fields. Now, then, compare these two factors: (1) the patterns drawn by magnetic fields, and (2) the patterns I'm saying the six heavenly points draw around the tertiary plane. Do the two look similar to you?

Some may object: "You're doing nothing more than to rotate things around three different axes, and that alone is supposed to place them in worlds so totally divided from one another that they can move at astronomical velocities in one without crashing into anything in the other?! That's unthinkable." To that, I reply: "That's the way it is with purely logical universes. Something as flimsy as a purely logical distinction can set things *worlds* apart from one another; even though, from God's viewpoint, all is contained within the same, one utterly non-spatial point.

H. ESOPTRICS ON INTERGALACTIC SPACE TRAVEL:

Let's close this document with a most remarkable idea suggested by the notion of these patterns of divine rotation in the primary and secondary planes. In addition to offering us an explanation of magnetic lines of force, they also give us something of an insight into how it is possible to go from virtually any one place in the universe to virtually any other place in the universe in what the occupants of our plane would call an astronomically insignificant fraction of a second.

To understand how that could be so, look again at figure #35 on page 157, and leave Earth where it is. From what we have said, Earth will—in the primary plane—swing from its depicted side of the Sun all the way to the opposite side and then back again in 10^{-73} seconds. From that it follows that, in one-half of that 10^{-73} seconds, Earth is very briefly on that side of the Sun which is directly opposite the one in which our figure places it. Ah! Do you see it? What if?! Yes! What if, in the course of the 10^{-94} seconds that it's in that opposite location, the Earth (or some vehicle on Earth) could ***drop out of the rotation in the primary plane and revert to the tertiary plane***? Could that vehicle do such a thing, it would—*in a mere 10^{-73} seconds by **our** reckoning!*—move from one point in the Earth's tertiary orbit around the Sun to a second point which is on the other side of the Sun directly opposite the first one.

That, though, is only a minor feat. The Sun's carrying generator is at the outer limits of a form whose center is at the center of the galaxy, and, in the course of ***its*** rotation in the primary plane, ***that*** form takes the center of our solar system from one side of the galaxy to another in an astronomical fraction of what we call a second. But, the galaxy, too, is subject to a form whose center is at the center of a super galaxy, and, in the course of ***its*** rotation in the primary plane, ***that*** form takes the center of our galaxy from one side of its super galaxy to another in an astronomical fraction of what we call a second. But, the super galaxy, too, is subject to a form whose center is at the center of a cluster of super galaxies, and, in the course of ***its*** rotation in the primary plane, ***that*** form takes the center of the super galaxy from one side of its cluster to another, and so forth. Eventually, there is a form whose center is near the center of the universe, and, in ***its*** rotation in the primary plane, ***that*** form takes clusters of clusters of super galaxies from one side of the universe to another in an astronomical fraction of what we call a second. To go from any one point in the universe to any other, it's merely a question of knowing how to drop out of which rotation when and where. Can you even begin to imagine what it must be like to have a level of technology which allows you to do that?

The synthetic propositions that make up the content of the sciences can be known only on the basis of experience; science depends ultimately, and in a fundamental way, on observation. This is the *empiricism* of analytic philosophy, central to the attack on traditional metaphysics, carried on, for instance, by Reichenbach. In his interpretation, metaphysicians like Kant took for granted the existence of synthetic *a priori* propositions—that is, propositions which can be known independently of experience, but which nevertheless are not mere tautologies. The task of the metaphysician was only to discover which they are, and to explain how it is that we can come to know them. It is this conception, Reichenbach argued, that gave philosophy a deservedly bad name among scientists, for the philosopher pretended that his armchair speculations, if sufficiently profound, could answer questions about the nature of things, while the scientist, asking much more modest questions, was constrained to go into his laboratory for experiment and observation.

——**ABRAHAM KAPLAN:** *The New World Of Philosophy* as given on page 81 of the Vintage Book published by Alfred A. Knopf, Inc. and Random House, Inc., New York. ©1961 by Abraham Kaplan.[1]

---

[1]In the above, when Mr. Kaplan speaks of "experience", I dare say he's talking about the kind in which one supposedly observes the extra-mental world. If not, then, when he speaks of "propositions which can be known independently of experience", we'd have to imagine he's talking about propositions which can be known without any kind of awareness of anything whatsoever. Since that's too ludicrous to be possible, it must be he's using the term "experience" in a way which leaves it by no means synonymous with "awareness" and "consciousness", but synonymous with the phrase "the scientist's act of observing the extra-mental world". If still alive, what will he think when he sees what this rank amateur uncovers with nothing more than the "armchair speculations" of a would-be mystical philosopher?

Document #8:

THE SIMILARITY OF SEXTONS & SUB-SEXTONS TO QUARKS & LEPTONS:

A. REVIEW & AMPLIFY:

In the course of document #7, pages 151 thru 156, it was said that there are 6 species of sextons distinguished from one another by the *categorical* OD to which each is native. One species is native to $OD2^4$, a second to $OD2^8$, a third to $OD2^{16}$, a fourth to $OD2^{32}$, a fifth to $OD2^{64}$, and a sixth to $OD2^{128}$. Each species is then either positive or negative depending upon which way its rotation in the tertiary plane takes it around the B'C' axis. If its rotation around B'C' involves A→A'→B→C, then I arbitrarily call its rotation positive. If its rotation around B'C' involves A→C→B→A', then that's most certainly the reverse of the former and, therefore, is an anti-rotation or negative rotation *relative to the other*. Of course, if you prefer to call the first sequence a negative rotation and the second a positive one, I don't propose to insist you are in error. For, I have yet to find any evidences whatsoever to indicate which rotation *demands* the label "positive" and which the label "negative". For now, to each his own! Whatever you decide on that point, we now have 12 sextons—1 positive and 1 negative at each of the 6 OD's 2^4, 2^8, 2^{16}, 2^{32}, 2^{64}, and 2^{128}.

In the course of document #7, pages 151 thru 156, it was also said that the sextons further break down into 3 sub-species distinguished from one another by which of the three axes serves as it tertiary plane of rotation. For one kind of sexton, positive and negative rotations in the tertiary plane are around the B'C' axis; for another, positive and negative rotations in the tertiary plane are around the A'C axis; and for a third kind, positive and negative rotations in the tertiary plane are around the AB axis. That now gives us a total of 36 sextons—3 positive and 3 negative at each of the 6 OD's 2^4, 2^8, 2^{16}, 2^{32}, 2^{64}, and 2^{128}. The chart on the next page spells that out for us graphically. In the chart, OD_n = the native OD of the given sexton, and A_3 = the axis of rotation in the tertiary plane.

At first glance, this seems to match exactly with what a scientist is soon to tell us about the quarks. Don't be fooled. The 36 kinds of sextons listed in chart #8 on the next page are not really 36 kinds of particles; rather, they are the 36 ingredients which go together to produce 72 kinds of sextons. More specifically, they go together to produce 6 positive and 6 negative sub-species of sextons each of which then applies to the 6 species of sextons which I've dubbed 2^4, 2^8, 2^{16}, 2^{32}, 2^{64}, and 2^{128}. As you can readily calculate, that's: 12 x 6 = 72.

OD_n	A_3	$A'{\to}B'{\to}$ $C{\to}C'$	$A'{\to}C'{\to}$ $C{\to}B'$	A_3	$A{\to}B'{\to}$ $B{\to}C'$	$A{\to}C'{\to}$ $B{\to}B'$	A_3	$A{\to}A'{\to}$ $B{\to}C$	$A{\to}C{\to}$ $B{\to}A'$
2^4	AB	+Aleph	-Aleph	A'C	+Beth	-Beth	B'C'	+Daleth	-Daleth
2^8	AB	+Aleph	-Aleph	A'C	+Beth	-Beth	B'C'	+Daleth	-Daleth
2^{16}	AB	+Aleph	-Aleph	A'C	+Beth	-Beth	B'C'	+Daleth	-Daleth
2^{32}	AB	+Aleph	-Aleph	A'C	+Beth	-Beth	B'C'	+Daleth	-Daleth
2^{64}	AB	+Aleph	-Aleph	A'C	+Beth	-Beth	B'C'	+Daleth	-Daleth
2^{128}	AB	+Aleph	-Aleph	A'C	+Beth	-Beth	B'C'	+Daleth	-Daleth

CHART #8: THE 36 SUB-SPECIES OF SEXTONS.

For monons and dions, only _one_ of the 3 axes of rotation serves as the axis of rotation in the tertiary plane. For sextons, each of the _three_ axes of rotation can serve as the tertiary plane's axis of rotation. As a result, the sextons have 3 ways to attain 2/3 neutrality and 3 ways to attain 1/3 neutrality, and each of those 6 can be either positive or negative. Charts #9 and #10 below illustrate how that works. In the charts, the symbol ± signifies that both the positive and negative versions of the given form are present together.

SUB-SPECIES	COMBO	RESULT
+2/3 Daleth	±Aleph & ±Beth & +Daleth	+2/3 Neutral around B'C'
+2/3 Beth	±Daleth & ±Aleph & +Beth	+2/3 Neutral around A'C
+2/3 Aleph	±Beth & ±Daleth & +Aleph	+2/3 Neutral around AB
-2/3 Daleth	±Aleph & ±Beth & -Daleth	-2/3 Neutral around B'C'
-2/3 Beth	±Daleth & ±Aleph & -Beth	-2/3 Neutral around A'C
-2/3 Aleph	±Beth & ±Daleth & -Aleph	-2/3 Neutral around AB

CHART #9: 3 POSITIVE & 3 NEGATIVE SEXTONS EACH 2/3 NEUTRAL.

SUB-SPECIES	COMBO	RESULT
+1/3 Daleth	±Aleph & +Beth & +Daleth	+1/3 Neutral around B'C' & A'C
+1/3 Beth	±Daleth & +Aleph & +Beth	+1/3 Neutral around A'C & AB
+1/3 Aleph	±Beth & +Daleth & +Aleph	+1/3 Neutral around AB & B'C'
-1/3 Daleth	±Aleph & -Beth & -Daleth	-1/3 Neutral around B'C' & A'C
-1/3 Beth	±Daleth & -Aleph & -Beth	-1/3 Neutral around A'C & AB
-1/3 Aleph	±Beth & -Daleth & -Aleph	-1/3 Neutral around AB & B'C'

CHART #10: 3 POSITIVE & 3 NEGATIVE SEXTONS EACH 1/3 NEUTRAL.

As one can readily see, every sexton particle is a composite of either 4 or 5, logically concentric, leading forms each of which is native to the same categorical OD and each of which is at the same current OD. With every sub-sexton particle, the situation is very different from the standpoint of the leading form. Each has but a single leading form. In the case of the monons, that single leading form is the categorical one native to OD1. In the case of the two dions, either the leading form is the categorical one native to OD2, or it is the categorical one native to OD4. Because of that difference, all sub-sexton particles are 3/3 non-neutral; whereas, all sextons are either 2/3 or 1/3 neutral, which is to say either 1/3 or 2/3 non-neutral. Choose for yourself which way to state it.

For Aleph forms, their secondary plane of rotation is around the B'C' axis, and their primary plane of rotation is around the A'C axis. For Beth forms, the secondary plane is around AB and the primary around B'C'. That can be charted so:

TYPE	A₃	A₂	A₁
Aleph	AB	B'C'	A'C
Beth	A'C	AB	B'C'
Daleth	B'C'	A'C	AB

CHART #11: AXES OF ROTATION FOR ALEPH, BETH & DALETH SEXTONS.

I haven't yet mentioned the neutrons. Do I need to, or is it so obvious, my readers can figure out for themselves how Esoptrics would describe them? Only sextons can produce neutrons. Neutrons are the result of ±Aleph around the AB axis with a ±Beth around the A'C axis and a ±Daleth around the B'C' axis and all 6 of the same native OD and concentric at the same OD. In its internet article titled **Neutron**, Wikipedia says the neutron is composed of 3 quarks—one of which has a charge of +2/3 and 2 of which have a charge of –1/3. That would seem to throw out Esoptrics' version of a neutron, since Esoptrics' neutron is divisible into only 2 quarks like so: (1) ±Aleph around the AB axis with a +Beth around A'C and no rotation around B'C' thereby producing a positive 1/3 neutrality; and (2) ±Daleth around the B'C' axis and a -Beth around A'C and no rotation around AB thereby producing a negative 1/3 neutrality. That's one of the 3 ways to split Esoptrics' neutron into two quarks. The other 2 proceed so: In the second way, it's Aleph which is split in two, and, in the third, it's Daleth which is split in two. Why elaborate?

B. PROF. HAIM HARARI ON QUARKS & LEPTONS:

What little I know about leptons and quarks comes from a single article I read in the April 1983 edition of the **Scientific American**, Vol. 248: #4. Written by Prof. Haim Harari, the article was entitled: **The Structure Of Quarks And Leptons**. In the right-hand column of page 57 of the cited article, Prof. Harari first tells us that leptons and quarks are standardly given as the ultimate constituents of matter, which is to say they cannot be divided into more basic units. Obviously, Esoptrics disagrees and says each is composed of neutrinos and/or anti-neutrinos.

Turning to the leptons first, he lists 6: the electron, the electron type neutrino, the muon, the muon type neutrino, the tau, and the tau-type neutrino. Each of the three kinds of neutrinos has no electrical charge; but, -1 is the electrical charge for each of the other three leptons.

ESOPTRICS: LOGIC OF THE MIRROR

Turning next to the quarks, he again lists 6: bottom (*a/k/a: b*), charmed (*a/k/a: c*), down (*a/k/a: d*) strange (*a/k/a: s*), top (*a/k/a: t*), and up (*a/k/a: u*). He says the quarks *c*, *t*, and *u* each exhibit an electrical charge of +2/3; whereas, each of the other 3 (*i.e.: b*, *d*, and *s*) exhibit a charge of -1/3. Additionally, each quark may exhibit one of 3 kinds of color which Prof. Harari dubs, blue, red, and yellow. In the lower *left*-hand column of page 57, he warns us these colors have nothing to do with what we usually call colors. It's just the name given to one of the 3 long-range forces. Returning to the lower *right*-hand column of page 57, Prof. Harari tells us this: If each quark of a certain color is considered a distinct kind of quark, then 3 colors times 6 quarks gives us a grand total of 18 kinds of quarks. He gives special note to the fact that, though each of the quarks has both a color and an electrical charge, none of the leptons has color. He then adds that there is an anti-particle for each of the leptons and quarks already described. Each particle and its anti-particle have the same mass; but, their colors and electrical charges are reversed. For example, the positron is the anti-particle of the electron, and, whereas the electrical charge of the latter is *minus* 1, the electrical charge of the former is *plus* 1. By the same token, the red *up* quark has an electrical charge of *plus* 2/3; whereas, its opposite (*i.e.:* the antired *anti-up* quark) has a charge of *minus* 2/3. If, as he says, there are 18 quarks each with an anti-particle, then 18 quarks + 18 anti-quarks = 36.

Switch now to the lower part of the right-hand column on page 60 of the said article. There. Prof. Harari puzzles over a very intriguing pattern of electrical charges found in the leptons and quarks—namely: Going from -1 to +1 in intervals of 1/3, we find some particle exhibits one of the resulting charges. Why are there not other values such as +4/3 and -5/3? Then too, every particle which has an *integral* charge (*ex. gr.:* +1 or -1) has no color; whereas, every particle which has a *fractional* charge has color. He puzzles over whether that implies a connection between electrical charge and color and, therefore, between leptons and quarks.

Note what Prof. Harari gives us as the two main differences between the leptons and the quarks: On the one hand, the quarks have *fractional* electrical charges of 2/3 and 1/3; whereas, the leptons have only *non*-fractional electrical charges of 1. On the other hand, the quarks have color and the leptons do not. Then, too, he puzzles over whether there might be some connection between the electrical charge of a particle and its color.

Let us make two assumptions: (1) that what science calls an electrical charge corresponds to what I have called the various states of non-neutrality; and (2) that what science calls color corresponds to what I have called the 3 ways to rotate in the tertiary plane. If it be proper to make those two assumptions, then there is a rather startling similarity between what Prof. Harari says about leptons versus quarks and what Esoptrics says about sub-sextons versus sextons. What is even more startling is that Esoptrics does in fact delineate a very clear relation between the electrical charge and color. For, it explains in complete detail how the color found in the sextons is precisely what allows them to produce the *fractional* charges.

Perhaps the *most* astonishing point of all, though, is this: Prof. Harari wonders why nature has favored the charges of 1, 1/3, and 2/3. Esoptrics spells out a possible answer in complete detail right down to how an intrinsically triune infinity causes it all.

So much for what's similar! Let's now be candid about the differences.

The most obvious one is that, for Prof. Harari, there are 12 quarks but 72 for Esoptrics. That difference comes about because, for Esoptrics, the 6 positive and 6 negative quarks are variations which apply to each of the 6 sextons distinguished from one another by their native OD. I need hardly point out that Science has never heard of ontological distance, and it remains to be seen if Science could detect any difference between: (1) a sexton particle whose leading form is native to $OD2^{128}$, and (2) one whose leading form is native to some other OD. Naturally, if one can detect no such difference, then one must necessarily be left with the conclusion that there are only 6 positive and 6 negative quarks exhibiting 3 "colors". Hopefully, Science shall one day so dramatically improve its observational abilities as to discover that there are indeed 72 rather than 12 quarks. If and when that comes to pass, let them bear one thought in mind: Before millions of them with their Ph. D.'s and billions of dollars worth of equipment *observed* it, it—as with the presence of mass in the neutrinos and anti-neutrinos—was *deduced* by an amateur, mystical philosopher with only a high school education, an armchair, and a prayer for profound self-knowledge. How's that for a David and Goliath story?!

Here's another difference: If color has to do with 3 rather than 1 axis of rotation in the tertiary plane, then why do the sub-sextons not have one "color" instead of none? Don't they have 1 of the 3 axes which register with the scientists as "color"? I have no answer, save to say this: It possibly has something to do with the fact that sub-sextons have only one leading form; whereas, the sextons have 4 or 5 which relate to one another 3 ways whether in producing 2/3 or 1/3 neutrality. I'll have to let time solve this puzzle.[1]

There's also a difference regarding the leptons: Positive and negative muons and taus are perhaps the same as the two positive and negative dions native to OD2 and 4. But, what about the electron?! What is its anti-particle? Prof. Harari says the positron. I have no idea what its anti-particle might be according to Esoptrics, unless, as the monons accelerate beyond OD1, some rotate one way in the tertiary plane and others the opposite. Do all of one kind then orbit the atom's nucleus and all of the other kind nest inside that nucleus? For now, it's a repugnant thought. Hmm! On second thought, maybe the positron is a freak particle which endures only for a tiny fraction of a second when an electron is somehow briefly forced to rotate counter to its normal way. I just don't know.

I've already mentioned the neutrons. The discrepancy between what Science and Esoptrics say about them makes me strongly suspect I've missed something crucial. I say that because, from what Science tells me, neutrons and protons both seem to follow Esoptrics' most basic principle—namely: One factor and its opposite

---

[1] At 9:00 AM, CDT, on this Thursday, September 10, 2009, I have just finished reading an article on the Internet at en.Wikipedia.org/wiki/Quark%E2%80%93lepton_complementarity. It's dated January 2009 and, in part, refers to a "quark-lepton complementarity" which suggests that both leptons and quarks have 3 "colors". I do not see Esoptrics as absolutely ruling out the idea of Aleph, Beth, and Daleth sub-species at every OD. I only see it as absolutely ruling out the idea that those 3 sub-species could be made concentric below $OD2^4$.

seek neutrality thru a third called the nexus (*i.e.:* ☸). In the neutron, a –1/3 down quark serves as the nexus thru which a +/23 up quark and a -1/3 down quark successfully find neutrality. In the proton, a +2/3 up quark serves as the nexus thru which a +2/3 up quark and a –1/3 down quark unsuccessfully seek neutrality. Obviously, however far I've gone in 52 years, I still have a long way to go. If only there were someone else ready, willing, and able to complete the journey for me!

Finally, there are two discrepancies with regard to the neutrinos and anti-neutrinos. Prof. Harari lists only 3 neutrinos: (1) an electron type, (2) a muon type, and (3) the tau type. He then adds that, for each of these neutrinos, there is an anti-neutrino.

As for the first discrepancy here, Esoptrics says monons (*i.e.:* electrons?) all rotate in the same "direction" in the tertiary plane; and so, either every alephon emitted by them is a positive one (*i.e.:* neutrino?), or every alephon emitted by them is a negative one (*i.e.:* anti-neutrino?). Of necessity, then, there cannot be *both* an electron type *neutrino* and an electron type *anti*-neutrino. Between Science and Esoptrics, this is a discrepancy I cannot currently explain, unless I again fall back upon the above thought which, 2 paragraphs earlier, I declared repugnant to me.

As for the second discrepancy here, Esoptrics says neutrinos and anti-neutrinos are the ultimate constituents of the universe; and so, there must be an astronomical number of different kinds of neutrinos and anti-neutrinos. Why, then, does Science list only 3 kinds of each? Because of the pronounced difference between sextons and sub-sextons, I can easily imagine that alephons emitted by the latter would not, but alephons emitted by the former would, chain together to form different types of electromagnetic radiation including gamma rays and light. That would explain why only the neutrinos and anti-neutrinos emitted by sub-sextons are detectable as neutrinos and anti-neutrinos.

In conclusion, then, when it comes to the variety of ultimate building bricks, Prof. Harari and I are amazingly close. When it comes to the kinds of electrical charges and colors setting two basic divisions of particles apart from one another, we are at one. When it comes to explaining how those 2 differences arise and connect to one another, Esoptrics has a very clear answer and Science none.

Finally, in the middle of the middle column of page 60 of the said article, Prof. Harari notes how much triplication we find in nature. Three leptons have an electrical charge of -1. Three leptons are neutral. Three quarks have an electrical charge of +2/3, while its -1/3 in three others. For him, there is no reason for it. For Prof. Harari and his fellow scientists, there may indeed be no reason for the *triplication*. For Esoptrics, the reason is more obvious even than the nose on your face.

$\mathcal{D}$ocument #9:

ESOPTRICS' VIEW OF THE HYDROGEN ATOM:

A. A RECENT BUT PERHAPS MISTAKEN INSIGHT:

I greatly fear that everything said in this document is merely worthless nonsense able to do nothing but waste my and every reader's valuable time. Despite that fear, I shall leave this document as a part of this book anyhow.

For many decades now, I have been struggling in vain to discover much of what Esoptrics may be trying to tell me about the structure of the hydrogen atom. Yes, from early on, I knew it was telling me of: (1) a nucleus encompassing the ontological distances of 2 thru 2^{128}, and (2) a nucleus circumnavigated by a particle composed entirely of forms native to OD1. On August 21 of this year, 2009 (20 days ago as of this Sept. 10), the drawing below suddenly suggested to me a rather elaborate line of thought which, unfortunately, may be more worthless than the tits on a boar hog.

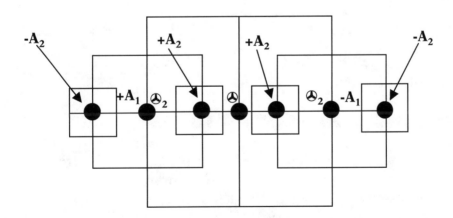

DRAWING #28 (REPEATED): GEOMETRICAL EXPRESSION OF ESOPTRICS' SECOND PRINCIPLE.

Suddenly, and perhaps mistakenly, I saw the above as suggesting this: Just as the great acceleration produces 8 reverse categories above the mirror threshold (*i.e.:* $OD2^{128}$), so also does it produce, below that threshold, 8 new, accelerated versions of the categorical OD's of 2^1, 2^2, 2^4, 2^8, 2^{16}, 2^{32}, 2^{64}, and 2^{128}. The result can be described as the 8 *regions* listed below in chart #12.

In line with the above drawing, #28, the first region contains 1 super-matrix (*i.e.:* accelerated form with its concentric partners) which serves as the anchor for the 7 subsequent regions. The second region contains 2 super-matrices; the third, 4; the fourth, 8; the fifth, 16; the sixth, 32; the seventh, 64; and the eighth, 128. Exactly how that works is spelled out in chart #13 below chart #12.

REGION	OD'S INCLUDED
#1	$2^{128}+1 \sim 2^{129}-1$
#2	$2^{127}+1 \sim 2^{128}-1$
#3	$2^{126}+1 \sim 2^{127}-1$
#4	$2^{125}+1 \sim 2^{126}-1$
#5	$2^{124}+1 \sim 2^{125}-1$
#6	$2^{123}+1 \sim 2^{124}-1$
#7	$2^{122}+1 \sim 2^{123}-1$
#8	$2^{121}+1 \sim 2^{122}-1$

CHART #12: 8 REGIONS OF THE HYDROGEN ATOM.

$\uparrow_A$	x	$\uparrow_B$	OD_n	#$\uparrow$	R
?	?	?	2^{128}	1	#1
2^{127}	2^{-1}	2^{126}	2^{1}	2	#2
2^{126}	2^{-2}	2^{124}	2^{2}	4	#3
2^{124}	2^{-3}	2^{121}	2^{4}	8	#4
2^{120}	2^{-4}	2^{116}	2^{8}	16	#5
2^{112}	2^{-5}	2^{107}	2^{16}	32	#6
2^{96}	2^{-6}	2^{90}	2^{32}	64	#7
2^{64}	2^{-7}	2^{57}	2^{64}	128	#8
TOTAL PARTICLES				255	

CHART #13: GREAT ACCELERATION BELOW THE MIRROR THRESHOLD.

In the first column, $\uparrow_A$ indicates the multiples involved in the great accelera-tion *above* the mirror threshold. In the second column, X indicates the multiple to be applied to column one in order to produce the number in the third column. In the third column, $\uparrow_B$ indicates the multiples involved in the great acceleration *below* the mirror threshold. In the fourth column, OD_n indicates the native OD of the form to be accelerated. In the fifth column, #$\uparrow$ indicates how many instances of the form mentioned in the fourth column are to be accelerated to the extent given in the third column. In the sixth column, R indicates to what OD the forms mentioned in the fourth column have been accelerated. For example, in the second row (not counting the header row), 2 forms native to $OD2^{1}$ have been accelerated to some-where between $OD2^{127}+1$ and $OD2^{128}-1$.

B. ELECTRON SHELLS:[1]

The above now suggests to me that the current OD of an accelerated monon (*i.e.:* electron?) can range anywhere from $2^{121}+1$ to $2^{128}-1$, thus giving the accelerated monons 7 atomic regions in which to play. Yes, there are 8 regions; but, no monon can access the 8th region. That's because no form can be accelerated more than 2^{128} times its native OD. Since every monon's native OD is 1, no monon can be accelerated beyond $OD2^{128}$. Furthermore, if accelerated to $OD2^{128}$, every monon races away at the speed of light, which is to say 1Φ per 2^{128}K.

Interestingly, the above concurs perfectly with Science's contention that there are 7 electron shells. According to Wikipedia's article ***Electron Shell***, Science calls them either K, L, M, N, O, P, and Q or 1, 2, 3, 4, 5, 6 and 7. Shell K (*i.e.:* #1) is the innermost shell; shell L (*i.e.:* #2) is one from the innermost shell; shell M (*i.e.:* #3) is two from the innermost shell; and so forth. Each shell then breaks down into anywhere from 1 to 7 sub-shells called s, p, d, f, g, h, and i. The breakdown is so:

SHELL	SUB-SHELLS PER SHELL
K	s only
L	s & p
M	s, p & d
N	s, p, d & f
O	s, p, d, f & g
P	s, p, d, f, g & h
Q	s, p, d, f, g, h & i

CHART #14: THE 7 SHELLS & 7 SUB-SHELLS.

Each of the sub-shells can house up to a definite number of electrons. Chart #15 below gives the maximum number of electrons per sub-shell. The left column gives the name of the sub-shell, and the right column gives the maximum number of electrons per sub-shell.

SUB-SHELL	ELECTRONS
s	2x1=2
p	2x3=6
d	2x5=10
f	2x7=14
g	2x9=18
h	2x11=22
i	2x13=26

CHART #15: MAXIMUM NUMBER OF ELECTRONS PER SUB-SHELL.

---

[1] Sept. 11, 2009, this insight came to me as I was writing this document. It perhaps justifies my decision to go ahead with this document no matter how worthless it seemed at first.

Because the K shell has only the s sub-shell, the K shell houses only 2 electrons. Because the L shell has both the s and the p sub-shells, the L shell houses 2 + 6 = 8 electrons. Because the M shell has both the s, the p, and the d sub-shells, the M shell houses 2 +6 +10 = 18 electrons, and so forth. Chart #16 below gives the whole picture in detail.

SHELL	MAX ELECTRONS PER SHELL =$2N^2$
K = #1	2
L = #2	2+6 = 8
M = #3	2+6+10 = 18
N = #4	2+6+10+14 = 32
O = #5	2+6+10+14+18 = 50
P = #6	2+6+10+14+18+22 = 72
Q = #7	2+6+10+14+18+22+26 = 98

CHART #16: MAXIMUM NUMBER OF ELECTRONS PER SHELL.

Look at chart #16 carefully, and you'll notice this: The maximum number of electrons per shell = $2(N^2)$ where N = the number of the shell. Does that formula sound familiar. It's one of Esoptrics most fundamental formulas, and it's one I've been well aware of for at least 48 years. See formula #5 on page 36.

SHELL # =	REGION #
1	8
2	7
3	6
4	5
5	4
6	3
7	2

CHART #17: WHICH SHELL IS ASSOCIATED WITH WHICH REGION.

Chart #12 on page 154 tells us: (1) that region #7 includes 2 times as many ontological distances as does region #8; (2) that region #6 includes 4 times as many OD's as does region #8; (3) region #5 includes 8 times as many; (4) region #4 includes 16 times as many; (5) region #3 includes 32 times as many; and (6) region #2 includes 64 times as many. Add to that what chart #17 tells us, and it makes sense to expect: (1) that shell #2 shall hold at least 2 times the number of electrons #1 holds; (2), that shell #3 shall hold at least 4 times the number of electrons #1 holds; (3) that shell #4 shall hold at least 8 times the number of electrons #1 holds; and (4) so forth. That's 2, 4, 8, 16, 32, 64, and 128 versus chart #16's 2, 8, 18, 32, 50, 72, and 98. It's successive powers of 2 versus 2 times the square of successive whole numbers. It's also interesting to note that, in drawing #15, we go from sub-shell to sub-shell by successive multiples of 4. Thus: 2+(1x4)=6; 2+(2x4)=10; 2+(3x4)=14; 2+(4x4)=18; 2+(5x4)=22; and 2+(6x4)=26.

Is there anything significant in those numbers? None I can see! I'm just throwing them out for the sake perhaps of aesthetics. For some of us, you see, numbers are among life's most beautiful experiences.

In its article **Electron shell**, Wikipedia says outer shell Electrons travel farther from the nucleus because the pull of the atom's nucleus upon them is weaker and, therefore, more easily broken. Esoptrics says how far an electron travels from the nucleus has nothing to do with any "pull" on the part of the atom's nucleus. It has to do with the fact that the electron's leading form causes its carrying generator to be in potency to whatever form has the current OD demanded by the current OD of the electron's leading form. The electron then stays where it is, because its carrying generator cannot go beyond the outer limits of the LB&D of the form to which the electron's leading form has limited its carrying generator. Then, too, bear in mind that, like all forms, the form to which the electron's carrying generator is in potency shall expand and contract its LB&D at regular intervals. That means the electron must alternately, and at regular intervals, dive toward the center of the nucleus and swing away from it. If memory serves me, that's been observed.

An electron in region #8 causes its carrying generator to be in potency to a form concentric with an accelerated form whose native OD is 2^{64}. An electron in region #7 causes its carrying generator to be in potency to a form concentric with an accelerated form whose native OD is 2^{32}. An electron in region #6 causes its carrying generator to be in potency to a form concentric with an accelerated form whose native OD is 2^{16}. No doubt, my readers can take it from there.

To go beyond its current distance from the nucleus, the electron must cause its carrying generator to become in potency to a form with a larger LB&D, which, in turn, means the electron must cause its carrying generator to become in potency to a form with a current OD which is higher than the OD of the form which held the electron closer to the nucleus. To have such an effect, the electron's leading form must increase its own current OD. For each increase in its current OD, the electron's leading form shall ingest one neutrino (anti-neutrino?), and vice versa.

C. DIFFICULTIES ASSOCIATED WITH ATTEMPTS TO RECONCILE CHART #13 WITH DRAWING #28:

Apparently, the insight expressed in charts #12 and #13 has apparently led me to uncover a startling similarity between what Science and Esoptrics say about electron shells. That seems to indicate rather strongly that there is something very correct about that insight. For all of that, I find myself most manifestly confronted with 2 grave difficulties when I try to reconcile charts 12 and 13 with drawing #28.

The first difficulty obvious to me is the nature of the single particle in region #1. I have long looked upon it as having a native OD of 2^{128} and a current OD of somewhere between $2^{128}+1$ and $2^{129}-1$. Also, I have long looked upon it as being the leading form of a super matrix in which at least $2^{128}-1$ forms are concentric. The –1, of course, represents the idea that the form native to OD1 cannot be captured by any but another OD1 form. How, now, does it manage to capture (*i.e.:* make concentric to itself) those other 2^{128}-2 forms? Chart #13 on page 178 doesn't give a clue.

As for the second difficulty obvious to me, it is brought home to me by what Science says is the composition of the proton. How is that? If I apply chart #13 to drawing #28, then the +A_1 and the –A_1 are each a lepton, because, according to chart #13, each is a super matrix in which the leading form is one native to OD2. Science, though, says each must—like ⊗ in drawing #28—be a quark.

After mulling on this for the last few days, I have this day, Saturday, September 12, 2009 (feast of the Holy Name of Mary)[1] come up with a solution which may, as they say, "kill two birds with one stone". What is that solution's nature?

I have long toyed with the idea that, when the great eight-fold acceleration took place _above_ the mirror threshold, it produced a _reverse_ effect _below_ that threshold. And what was the nature of that reverse effect? Without either accelerating or decelerating, forms native to OD2^{128} made concentric to themselves one form at each of the ontological distances of 2^1 thru and including 2^{128}-1. In other words, as—in the first stage of the great eight-fold acceleration—a form native to 2^{128} accelerated to somewhere between OD2^{255}+1 and OD2^{256}-1, the automatic, reverse result was a super matrix composed of one form at OD2, one at OD3, one at OD4, one at OD5, etc., all the way to, and including, the leading form at OD2^{128}+1 at least. How far the super matrix goes beyond OD2^{128}+1, depends upon the velocity of the super matrix. See the formula for velocity (_i.e.:_ $S_{AG}[C/N_G]$), on page 47.

I should have immediately seen what that automatically implies about the other stages of the great eight-fold acceleration; but, I did not begin to see it until late yesterday, and, only a little earlier this morning (It's now 10:00 AM, CST), did I see it quite clearly. As for why, I guess I was always too fascinated with the categorical OD's to pay much attention to the generic ones. Here, then, is what is now suggested to me as a solution to the two problems we're now discussing:

X↓	PROGRESSION	OD$_n$	#P↓	OD$_c$	#Matrices/Particle
2^{128}-1	X = 2^{128}	2^{128}	1	2^{128}+1 min.	OD2~OD2^{128}+1 min.
2^{127}-1	2^{-1}x = 2^{127}	2^{127}	2	2^{127}+1 min.	OD2~OD2^{127}+1 min.
2^{126}-1	2^{-2}x = 2^{126}	2^{126}	4	2^{126}+1 min.	OD2~OD2^{126}+1 min.
2^{124}	2^{-4}x = 2^{124}	2^{125}	8	2^{125}+1 min.	OD2~OD2^{125}+1 min.
2^{120}	2^{-8}x = 2^{120}	2^{124}	16	2^{124}+1 min.	OD2^4~OD2^{124}+1 min.
2^{112}	2^{-16}x = 2^{112}	2^{123}	32	2^{123}+1 min.	OD2^{11}~OD2^{123}+1 min.
2^{96}	2^{-32}x = 2^{26}	2^{122}	64	2^{122}+1 min.	OD2^{26}~OD2^{122}+1 min.
2^{64}	2^{-64}x = 2^{64}	2^{121}	128	2^{121}+1 min.	OD2^{57}~OD2^{121}+1 min.
TOTAL PARTICLES			255		

CHART #18: REVISED VERSION OF CHART #13.

In the above chart, X↓ gives us the multiple of the downward fusion. The second column from the left is, I surmise, self-explanatory. OD$_n$ gives us the native OD of the form to which X↓ is applied. The column marked #P↓ gives us the number of

---

[1] If the insight I'm about to spell out proves to be correct, then let our Dear Lady's Name be praised.

particles produced per row by the downward fusion. For example, only 1 instance of the particle whose leading form is native to $OD2^{128}$ is produced by the downward fusion; 2 instances of the particle whose leading form is native to $OD2^{127}$ are produced; 4 instances of the particle whose leading form is native to $OD2^{126}$ are produced; and so forth. OD_c gives us the minimum current OD of the form produced by $X\downarrow$. I cannot give the *exact* OD_c, because that depends upon the current velocity of the particle in question. Finally, the column marked #Matrices/Particle tells us what forms are to be found concentric in each of the particles produced in that row. For example, in the 5th row down (not counting the header row), each of the 16 particles contains 2^{120} forms concentric with one another.

If we apply this revised chart to drawing #28, then we find both of the prior difficulties solved. For, on the one hand, we now know how the main form represented by ⊛ manages both to have the current OD it does have and to be composed of 2^{128} forms concentric with one another. On the other hand, we now see that $+A_1$ and $-A_1$ are quarks. One is positive because of the way it rotates, and the other is negative because of the reverse way it rotates.

Unfortunately, when it comes to the 3 particles making up the proton, Science say 2 of them are +2/3 quarks and one is a -1/3 quark. But then, what amounts to a $+A_1$ versus a $-A_1$? Is it a case of a +1/3 versus a –1/3, or is it a +2/3 versus a –1/3? Someone smarter than I shall have to answer that question.

Well, on second thought, maybe I can answer it myself. One must bear in mind that drawing #28 is Geometry speak's way of diagramming the *successful* pursuit of neutrality. That makes it reasonable to assume there are other diagrams of the *un*successful pursuit of neutrality. Indeed, in Science's description of the neutron, it appears there may be more than one way to diagram the *successful* pursuit of neutrality.

There's also this question waiting to be answered: According to chart #18, there are no particles whose leading form is native either to OD2 or OD4. Where, then, are the 2 kinds of leptons called muons and taus? Are they merely freak particles momentarily created by busting up the atom? And what of the positrons? Are they the same sort of product? As I say, someone smarter than I am shall have to answer those questions.

For now, all that matters to myself is this: Whatever the questions left unanswered, I dare say Esoptrics has—to Science's cosmological model—come close enough to achieve its main goal—namely: to leave every honest thinker confessing: "We can no longer be certain that physical extension and infinite divisibility are **_not_** just 3 more colossal cosmological blunders."

The experience from which the empiric draws his conjectures is, of course, the homely and substantial experience of a world of public objects, which forms for all sane and unreflective persons the basis of ordinary life. It has been regularly insisted, however, since the earliest times, that experience in this sense is nothing ultimate: the alleged paradoxes of motion and change and the more familiar facts of perceptual error and illusion are enough (it is thought) to show that it cannot be straightforwardly identified with the real. Hence, in addition to the rejection of habit-learning as a road to knowledge, there arises that further prejudice against the deliverances of the senses and in favor of necessary reasoning from first principles, of which the Parmenidean distinction between the "ways" of truth and opinion is an early and famous example.

——**P. L. HEATH:** *Experience*, as given on page 157 of the article in volume 3 of the 1972 reprint edition of ***The Encyclopedia Of Philosophy*** published by the Macmillan Publishing Co., Inc. & The Free Press. © 1967 by Macmillan, Inc..

Document #10:

E-MAIL OF MARCH 25, 2007:[1]

A. SIR MICHEAL ATIYAH ON RELATIVITY VS. QUANTITUM PHYSICS:

Many thanks to you, Alex,[2] for the copy of Sir Michael Atiyah's lecture on mathematics in the 20th century.[3] Upon receiving the e-mail version of Sir Michael's lecture, the first thing I did was to convert it into a doc file. I then removed the superabundant supply of paragraph marks, inserted what few I thought were appropriate, and then inserted all the instances of "fi" and "ff" which, for some unknown reason, had apparently become lost in transmission. Why only those letters?!

I did all of that because, I am loathe to read anything, unless I can mark the passages I find interesting, and, that, I could not have done had I been content to read the original transmission. As you probably know, one cannot edit a received e-mail message. Well, at least I do not know how to do that without converting it.

Then too, I wanted to be able to use the search utility which works so well once one converts the e-mail to a doc file. With that search utility available, I could then double check questionable points. For example, in your heading you wrote: "this paper by Atiyah discusses where the mismatch between special relativity and quantum physics comes from". Using the search utility, I find the term "special relativity" occurs nowhere in Sir Michael's article. Indeed, the term "relativity" occurs only twice in the entire document. First, it occurs on what the document itself describes as page 9. There, we read:

> That was the theory developed by Elie Cartan, the basis of modern differential geometry, and it was also the framework that was essential to Einstein's theory of relativity. Einstein's theory, of course, gave a big boost to the whole development of differential geometry.

Secondly, it occurs on what the document itself describes as page 11. There,

---

[1] This is a highly edited version of the original.

[2] At the time, Alex described himself only as "a scientist at one of the well known universities". He was one of the few individuals ever to show any interest in Esoptrics. As usual, his interest did not last long.

[3] This lecture was apparently delivered as part of the Fields Lectures at the World Mathematical Year 2000 Symposium, at Toronto, Canada, on June 7th thru the 9th, 2000.

one reads:

> I have already mentioned general relativity and Einstein's work.

It is to *general* relativity that Sir Michael refers and not to *special* relativity.
Conversely, Sir Michael uses the term "quantum" some 23 times in such combinations as "quantum mechanics", "quantum theory", "quantum field", "quantum groups" (which, on pg. 13, he says, are not groups at all), "quantum cohomology", and "quantum mathematics". Nowhere, though, do I find him making a specific statement such as: "The difference between Relativity and Quantum Mechanics is such-and-such." That, unfortunately, leaves me somewhat in the dark. Reading between the lines, though, I find him repeatedly making a connection between Quantum Theory and "infinite-dimensional space", and, however repugnant to me, he seems to make a big deal of "infinite-dimensional space". For example, on page 12, he writes:

> As I mentioned before, the 20th century saw a shift in the number of dimensions ending up with an infinite number. Physicists have gone beyond that. In quantum field theory they are really trying to make a very detailed study of infinite-dimensional space in depth. The infinite-dimensional spaces they deal with are typically function spaces of various kinds. They are very complicated, not only because they are infinite-dimensional, but they have complicated algebra and geometry and topology as well, and there are large Lie groups around, infinite-dimensional Lie groups.
>
> So, just as large parts of 20th-century mathematics were concerned with the development of geometry, topology, algebra, and analysis on finite-dimensional Lie groups and manifolds, this part of physics is concerned with the analogous treatments in infinite dimensions, and of course it is a vastly different story, but it has enormous payoffs.

On pages 13 and 14, he adds:

> What about the 21st century? I have said the 21st century might be the era of quantum mathematics or, if you like, of infinite-dimensional mathematics.

Now, as I understand what Einstein supposedly said, one deals only with four dimensions, three of which are space and the fourth is time. Of course, he so often speaks of space-time, it seems he would not speak of three dimensions of space and 1 of time and would prefer to speak of 4 dimensional space-time. I, of course, speak of space itself as four dimensional or, as I would prefer to put it: Every matrix is three dimensional; but, they wind up stacked up inside of one another somewhat like three-dimensional boxes inside of increasingly larger three-dimensional boxes and, thereby, produce a fourth dimension.

In short, then, it seems Sir Michael is saying the difference between Relativity and Quantum Theory is four-dimensional time-space versus infinite-dimensional mathematics. What do you think? Is that a correct assessment on my part?

B. SIR MICHAEL ON ALGEBRA VS. GEOMETRY:

Sir Michael really got to me on one issue, because of how important it is to me. On pages 4 thru 7 of his lecture, he talks about Geometry versus Algebra and "algebraists" versus "geometers". Geometers, he says, are Newtonians and try to explain the world around us in the terms of Geometry and geometrical figures, and this, he says, means in the terms of space. We know only too well, of course, what Newton meant by space. Algebraists, he says, follow Leibniz in choosing Algebra over Geometry. His own words are:

> Algebra is concerned with manipulation in time, and geometry is concerned with space. These are two orthogonal aspects[1] of the world, and they represent two different points of view in mathematics. Thus the argument or dialogue between mathematicians in the past about the relative importance of geometry and algebra represents something very fundamental. - Page 6.

> If you compare in analysis the work of Newton and Leibniz, they belong to different traditions: Newton was fundamentally a geometer, Leibniz was fundamentally an algebraist, and there were good, profound reasons for that. For Newton, geometry, or the calculus as he developed it, was the mathematical attempt to describe the laws of nature. He was concerned with physics in a broad sense, and physics took place in the world of geometry. If you wanted to understand how things worked, you thought in terms of the physical world, you thought in terms of geometrical pictures. – Pages 4 & 5.

Yes, I, too, however much I prefer Algebra to Geometry, recognize the need to resort to Geometry's pictures. How broadly I must smile in agreement, then, when I read Sir Michael saying:

> This choice between geometry and algebra has led to hybrids that confuse the two, and the division between algebra and geometry is not as straightforward and naive as I just said. For example, algebraists frequently will use diagrams. What is a diagram except a concession to geometrical intuition? – Page 7

---

[1] Orthogonal aspects???!!! What in God's name is that supposed to mean? According to Webster, "orthogonal" means what's related to, or composed of, right angles. Is he saying Algebra and Geometry are the two aspects of the world related to, or composed of, right angles???!!! Perhaps he's trying to say Algebra and Geometry are the 2 *cornerstones* of the world of mathematics. Cornerstones are composed of right angles, are they not? If that's what he meant, then pray tell me for sanity's sake why anyone would say "2 orthogonal aspects" when one could say "2 cornerstones".

Understanding, and making sense of, the world that we see is a very important part of our evolution. Therefore spatial intuition or spatial perception is an enormously powerful tool, and that is why geometry is actually such a powerful part of mathematics not only for things that are obviously geometrical, but even for things that are not. We try to put them into geometrical form because that enables us to use our intuition. – Page 5.

In a very humorous vein, he tells us about what he calls the "Faustian Offer". Algebra, he says, is what the devil offers Faust in exchange for his soul. His own words are:

One way to put the dichotomy in a more philosophical or literary framework is to say that algebra is to the geometer what you might call the "Faustian Offer". As you know, Faust in Goethe's story was offered whatever he wanted (in his case the love of a beautiful woman) by the devil in return for selling his soul. Algebra is the offer made by the devil to the mathematician. The devil says: "I will give you this powerful machine, and it will answer any question you like. All you need to do is give me your soul: give up geometry and you will have this marvelous machine." [Nowadays you can think of it as a computer!] – Page 6.

Sir Michael's predilection for Geometry is further illustrated by his reference to "Arnold" who, I assume, is Thomas Arnold 1795-1842. Arnold, says Sir Michael:

. . . . makes no bones about the fact that his view of mechanics, in fact, of physics, is that it is fundamentally geometrical, going back to Newton; – Page 5.

Well, I'll drop Arnold and Newton like hot potatoes, take the "Faustian Offer" any day of the week, and make no bones about the fact that my view of mechanics, in fact of physics, is that it is fundamentally algebraical. When the geometers try to reduce the universe to what's physical, describable by geometry, and made separate or overlapping by _space_, I say it's all self-contradictory gibberish. Because space is ultimately nothing like what the geometers imagine it to be, the geometers—though they see it not—leave nothing in this universe either separate or overlapping. On the contrary, the viewpoint of geometers like Newton and Arnold leaves everything in the universe always, wholly, entirely, and perfectly coincidental with, and indistinguishable from, everything else in the universe, because space is simply not what Geometry fancies it is.

The opposite is true of the viewpoint given us by Algebra and algebraic logic. Its viewpoint gives us true separation and overlapping, because all separation and overlapping is, in reality, purely logical and algebraical rather than spatial and geometrical, and that's why two particles light years apart according to the geometers can, under some circumstances, instantly communicate with one another without recourse to communication over the vast distances so dear to the geometers. Those vast distances simply aren't there the way Geometry's spatial fantasies imagine them there, and the geometers are merely "smoking rope" when they appeal to those vast distances.

Sir Michael has it backwards when he talks about Algebra being what the devil offers to Faust in exchange for his soul. It is Geometry which the devil offers to Faust. On what grounds do I say that? Well, where you have a universe describable in the terms of Geometry's pictures, you have a universe which can be described without recourse to the concept of God as an Infinitely Informed Intelligence. And isn't it the devil's great ambition to remove God from the picture?! Where you have a universe describable by Geometry, you have one which doesn't need the *real* God. That's because it has the most miraculous *"Deus ex machina"* every lowered onto a stage—in this case, ever lowered onto the stage of a farce.

I refer, of course, to space—to space as the kind describable by Geometry, which is to say the kind that, in Geometry's pictures of it, is little or no different from the picture we get from our senses. Geometry's space certainly is the most miraculous pseudo-god imaginable. After all, nothing could be more miraculous than a ball of pure, self-sustaining nothingness which—despite being pure nothingness—nevertheless has at least finite, if not infinite, length, breadth, and depth. How pure nothingness could be that real is utterly beyond me. (Is there something revealing in this business of great minds rejecting the infinite God Who is intelligent but accepting the infinite pseudo-god who is pure nothingness?)

In the opposite direction, where you have a universe describable in the terms of Algebra, you have a picture of space *radically* different from the picture our senses give us. For, Algebra's "picture" of space is a 3-dimensional graph showing us how many potential and actual *primary* acts are where in a logical order of *sequence*. Without an infinitely informed overseer to author and enforce that logical order of *sequence*, there is no such thing as Algebra. In the end, Geometry is basically a static, photographic reproduction of our visual images; whereas, Algebra is a graph of what kind of logical sequences can be found among the most fundamental operations in our universe. As Sir Michael expresses it in his lecture:

> Algebra, on the other hand (and you may not have thought about it like this), is concerned essentially with time. Whatever kind of algebra you are doing, a *sequence of operations* [Emphasis mine – ENH} is performed one after the other, and "one after the other" means you have got to have time. In a static universe you cannot imagine algebra, but geometry is essentially static. I can just sit here and see, and nothing may change, but I can still see. Algebra, however, is concerned with time, because *you have operations that are performed sequentially* [Emphasis mine – ENH] and, when I say "algebra", I do not just mean modern algebra. – Page 6.

Yes, Geometry is basically a static picture of what is where in spooky space, and, as long as space is the self-subsisting reality it appears to be, Geometry has no need for God in its pictures. Algebra, though—especially as Esoptrics—is a "picture" of which acts are where in the order of sequence. Needless to say, an order of sequence is not any kind of reality, let alone a self-subsisting one. It is merely a regimen, and a regimen automatically encourages us to look for its Author and Enforcer. Thus does Algebra point us toward God. I, therefore, shall ever proclaim Geometry the "Faustian Offer" and declare Algebra God's own divine logic.

C. SIR MICHAEL ON REJECTION:

Finally, there was one aspect of Sir Michael's lecture which I found particularly delightful. I refer to the way he mentions quite a few theories which he then classifies as having a "long way to go yet" [pg. 14]. I was maximally thrilled by this on page 10:

> Some years ago I was interviewed, when the finite simple group story was just about finished, and I was asked what I thought about it. I was rash enough to say I did not think it was so important. My reason was that the classification of finite simple groups told us that most simple groups were the ones we knew, and there was a list of a few exceptions. In some sense that closed the field, it did not open things up. When things get closed down instead of getting opened up, I do not get so excited, but of course a lot of my friends who work in this area were very, very cross. I had to wear a sort of bulletproof vest after that!

Yes, others can get multitudes in the academic world to listen at length to theories which have a long way to go yet and which are so lame they can't survive without their promoters threatening to shoot their opponents; but, Edward N. Haas can't get a single, solitary person to listen to his theory for long. How strange!

> When I first announced the results of my calculations at a conference at the Rutherford-Appleton Laboratory near Oxford, I was greeted with general incredulity. At the end of my talk the chairman of the session, John G. Taylor from Kings College, London, claimed it was all nonsense. He even wrote a paper to that effect.
> ——STEPHEN HAWKING: *A Brief History Of Time*, as found on pages 115 (bottom) and 116 (top) of the paperback, updated and expanded, tenth anniversary edition published by Bantam Books, New York, 1998. © 1988, 1996 by Stephen Hawking.

Well, Prof. Hawking, when one of your stature provokes such rejection, maybe I should not complain to any extent about what rejection I experience.

Document #11:

MORE EXCHANGES WITH MY BROTHER GORDON:

A. GORDON'S E-MAIL TO ME DATED AUGUST 25, 2009:

Ed, on a more detailed reply, I think you may want to add the third grand mistake made by way of our often faulty senses. That one being the basis for Newtonian Physics—the absolute space and time, which of course is what Einstein disassembled. The reason why I say you may want to include this is it would improve "the ramp" to Esoptrics because Relativity provided deep thought on just how "bizarre" and removed physical reality is from our senses. And, ultimately, this is what you are saying with Esoptrics.

Something to think about.

GORDON.

B. MY E-MAIL TO GORDON DATED AUGUST 26, 2009:

Precisely what Newton means by "absolute space" has recently become, for me, an open question. Until recently, it always seemed to me that Newton, like perhaps the vast majority of the human race, conceived of space as the void, and what most people mean by "the void" is that space is a giant ball of pure nothingness. As some express it: Space is a sheer vacuity of every kind of Being.

Some, such as Aristotle, rejected such a definition of space on the grounds that space would thus be nothing at all. That's what, at one stage in history, led to the conclusion that space is an immaterial kind of reality perhaps best called "the ether".

To my dismay, I recently came across writings on the Internet insisting Newton, too, accepted the notion that space cannot be pure nothingness but, rather is some kind of extension of God, whatever that means. Since I can make no sense of that, I cannot help but wonder precisely what absolute space was for Newton.

Then too, I recently came across a PDF on the Internet whose source is given only as nd.edu~Biener. Notre Dame University's web address is nd.edu; and so, I assume Biener, whoever he is, is at Notre Dame University. Disgustingly, when I went to Notre Dame's web site, there is no such person listed.

Anyhow, in his PDF posting, Biener writes:

ESOPTRICS: LOGIC OF THE MIRROR

I argue that Newton's notion of absolute, void space underwent a subtle change from the time of **De Gravitatione** to the 1720s. In particular, I argue that Newton started to question whether space was necessarily inert.

First of all, is Biener correct in writing "absolute, void space"? If so, then how do some say Newton did not accept the idea of space as a sheer vacuity of every kind of Being? Is Biener using "void" in some different sense such as: areas of space where nothing is present but space itself? If so, what is the nature of such space's reality? Next, if Newton's view of space changed toward the end, then to precisely what concept of space did he turn toward the end?

Actually, the truth of the matter is I'm not certain what any thinker, Einstein included, means by space. No philosopher or scientist I've ever read has ever given me an intelligible explanation of how space can be so permeable and devoid of any tactile characteristics, unless it is a vast expanse of pure nothingness. Esoptrics, on the other hand, gives a very detailed and mathematically precise explanation of: (1) what kind of reality space is (*i.e.:* the actuality of at least one form or, in most cases, of many forms logically concentric to one another), (2) why and exactly in what sense it is permeable, and (3) why neither human senses nor man made instruments can detect any of the internal characteristics of any form's primary act.

All of that, though, is a bit irrelevant. For, I would say that the idea of absolute space and time as a grand mistake is already addressed in Esoptrics' list of the 3 remaining "colossal cosmological blunders". For, absolute time is basically continuous time, and absolute space is infinitely divisible, physically extended space.

Your loving, exceedingly grateful brother:
EDWARD.

C. GORDON'S E-MAIL TO ME DATED AUGUST 26, 2009:

What I was referring to more so than the physical attributes of space and time (granted a deeper abstraction) was the mathematical characteristics as being independent variables. Under Newtonian physics, space and time cannot be functions of anything (*i.e.:* S = f[?]). Of course, under General Relativity, space is modified by mass, and time is modified by velocity—is what I was referring to. Also I was suggesting that Relativity more so than any other theory that preceded it proved that our senses can't be trusted to experience the validity of anything more abstract than bumping our big toe!

GORDON

D. MY E-MAIL TO GORDON DATED AUGUST 29, 2009:

D-1. OBJECTIVELY VS. SUBJECTIVELY RELATIVE TIME:

Greeting, my dear brother:

I assume that your formula, S = f[?], is saying that space (space-time?) is a function of: We know not what. One can easily understand how that would apply to Newton. After all, for Newton, time and space are absolutes (And yet, Newton does refer to space as an "emanation of God". Does that make it a function of God?) And who does not know this?: Absolutes are, by definition, not a function of anything else.

As for Einstein, I'm somewhat familiar with the notion that, in Relativity, "Space is modified by mass, and time is modified by velocity". With no knowledge of calculus, though, I have no ability to make any sense of the formulas wherein Einstein explains the characteristics of space-time with mathematics so technical, as to account exactly for such things as the planet Mercury's rather erratic orbit around the Sun and the exact extent to which light's path in the vicinity of the Sun should be deflected.

For me, then there is no possibility of going into the mathematical characteristics of space and/or time to the extent that Einstein did. There's even less possibility of my finding the time to learn the kind of mathematics with which the Einsteins of this world deal—assuming I even have enough brain power to learn something that complicated.

Still, Esoptrics does give some of the mathematical characteristics of space and time as *dependent* variables. Mostly, though, it doesn't go far beyond generalized concepts regarding: (1) how what Esoptrics calls "*subjectively* relative time" is a function of most "piggyback forms" and their "carrying generators", (2) how what Esoptrics calls "*objectively* relative time" is a function of one particular type of form, and (3) how space is a function of what Esoptrics calls "piggyback forms".

Subjectively relative time pertains to the vast majority of forms and generators, and, for any given one of them, is relative to the rate at which that given one changes from *primay* act (defined later) to *primary* act. In turn, that rate is determined by a combination of the given form's current ontological distance (OD for short) and its native OD. Subjectively relative time—unlike objectively relative time—is thus *very variable* (a contrast only Esoptrics explains) and, in Esoptrics as for Einstein, it slows with increases in velocity. That's because, in Esoptrics, an increase in velocity means an increase in the current OD of the accelerating form, and an increase in current OD means a decrease in the rate of change from primary act to primary act on the part of the accelerating form. It also means, for the generator carrying that form, an *increase* in the rate of change and, thus, a speeding up on the part of time *as it applies to the generator*; but, I'll not go into that.

Objectively relative time pertains only to forms changing from *primary* act to *primary* act at what is the fastest rate possible throughout this the universe's ninth epoch. For Esoptrics, that fastest possible rate is what our sensation limited minds would currently call one change per $7.201789375 \times 10^{-96}$ seconds. One Alphakronon (1K for short) is the name Esoptrics gives to this fastest possible rate of 1 change per 10^{-95} sec., which is to say $1.388543802 \times 10^{95}$ changes per second. Objectively relative time is still relative, because it does not give us the *duration* of primary acts on the part of the form changing from primary act to primary act at the fastest possible rate. All it tells us is how much more frequently it so changes.

For Esoptrics, objectively relative time is *absolutely invariable*. To give an example of what that means, consider the so-called "paradox" of the astronaut who goes off into space in the year 2010, returns 20 years later by his reckoning, and finds it's now the year 2100 on Earth. From the standpoint of objectively relative time, there's no difference, because both his 20 years in space and Earth's 90 years involved the same, identical number of alphakronons. How can that be?

In Esoptrics, the formula for calculating objectively relative time is:

$$K = D_c \times D_n$$

In the formula, D_c = current ontological distance, D_n = native OD, and K = the number of alphakronons. Naturally, in some cases, current and native OD are the same. Where that's true the formula becomes:

$$K = (D_n)^2$$

According to Esoptrics, we live our lives at an OD of somewhere between 2^{155} and 2^{156}, and our forms—because their current OD basically equals their native OD—change from primary act to primary act at the square of approximately 2^{155}. As you can readily calculate, that amounts to one change per roughly 2^{310}K (*i.e.:* 2.0859×10^{93}K). What we call a second is, in our experience of it, extremely brief. In the experience of forms currently at, and native to, OD1, our so-called "fleeting second" would seem to last as long as what we would call roughly 10^{83} millennia.

What produces that difference? **First of all**, it's because consciousness is a series of individual primary acts on the part of some individual's leading form.

Secondly, it's because, as long as a given individual's leading form is performing the same, one primary act, that leading form undergoes no kind of internal change whatsoever and, therefore, is to no extent conscious of any kind of change whatsoever. Where there is no consciousness of any kind of change whatsoever, the leading form has no consciousness of the duration of its acts of consciousness, which is to say the leading form is to no extent conscious of the *objectively* relative duration of its primary acts. All the leading form is aware of is its own personal rate of change from primary act to primary act, which is to say the leading form is conscious only of its own *subjectively* relative time. As a result, what a conscious leading form calls a "fleeting second" is actually a definite number of acts of consciousness, and I calculate that number to be somewhere between 20 and 30 per second, and that's the way *subjectively* relative time always is for all conscious leading forms everywhere in the universe.

That brings us to the **third** part of what brings about the above difference—namely this: An individual at OD1 experiences 1 second of subjectively relative time (*i.e.:* has a leading form which performs 20 to 30 successive primary acts) in the course of 20 to 30 alphakronons—thus making his *subjectively* relative time = *objectively* relative time. Every one of us, though, experiences 1 second of subjectively relative time (*i.e.:* has a leading form which performs 20 to 30 successive primary acts) in the course of roughly 10^{95}K. Therefore, what we experience as a second (*i.e.:* 20 to 30 successive primary acts), an individual at OD1 experiences as an astronomical number of successive primary acts, and that's how the great difference comes about between: (1) those for whom 10^{95}K involves only a fleeting second of 20 to 30 successive primary acts, and (2) those for whom 10^{95}K involves an astronomical number of fleeting seconds each of 20 to 30 successive primary acts.

Let's return now to the astronaut for whom 20 years in space turns out to be 90 years on Earth. Perhaps now you can see why that difference is there. For the astronaut and every person on Earth, the quantity of *objectively* relative time is the same—namely: roughly 10^{95} x the number of seconds in 90 years. Since the number of seconds in a year = 31,556,925.9747, we multiply that by $1.388543802 \times 10^{95}$ (*i.e.:* K per sec.), and thus calculate that 1 year = 4.38158×10^{102}K. Multiply that by 90, and we calculate that, for both the astronaut and every individual on Earth, his 20 year and their 90 year journey took approximately 3.94×10^{104}K of *objectively* relative time. The difference between: (1) the astronaut's experience of 20 years of *subjectively* relative time, and (2) each earthling's experience of 90 years of *subjectively* relative time is this: Every earthling's leading form changed from primary act to primary act 4.5 times as frequently as did the astronaut's leading form. Can I make it any simpler? And who else offers even the slightest trace of an explanation of the difference between *highly variable* vs. *absolutely invariable* time?

D-2. PRIMARY ACTS:

When it comes to the difference between Esoptrics and all other cosmological theories, one of the most difficult (if not *the* most difficult) concepts to grasp is the concept of a *primary* act. Every *primary* act is an act in no sense known to us. Every act and activity with which we—whether by means of our senses or by any of our instruments—are familiar is merely some kind of *secondary* act or activity produced when forms or generators change from *primary* act to *primary* act. Just as a person blind from conception cannot imagine colored shapes, neither can any *finite* being imagine what it's like to come face to face with the internality of a *primary* act. That's difficult for most to understand because, when most try to think of the internality of an ultimate constituent of the universe, they think of the deadpan rigidity of a billiard ball. In reality, there is nothing deadpan or rigid about the internality of an ultimate constituent of the universe. Every ultimate's internality is, in a manner of speaking, alive with a vibrance we cannot begin to imagine. Every *primary* act is not merely something which an ultimate *does*; it is, rather, something which an ultimate **is**, and every iota of what its internality **does/is** is—throughout each *primary* act—an operation without the slightest trace of any kind of internal change whatsoever. Remember Little Black Sambo's tigers racing around the tree so fast, they became a yellow blur which looked like motionless butter; and yet, they were far being either butter or motionless. They were a new kind of motion unlike any kind with which we are familiar.

This side of infinity, all we can ever know about any given primary act is the fixed quantity of which fixed kinds of power are being utilized throughout the course of that given primary act. For Esoptrics, there are 3 kinds of power each of which has an anti-power for a total of 6 modes of power. There are thus 3 combos of power/anti-power which Esoptrics labels: (1) A vs. B, (2) A' vs. C, and (3) B' vs. C'. Each of these 6 modes can have any particular one of 2^{256} possible levels and will remain at that one, particular level throughout the course of a primary act. That's how there's no internal change at all throughout the course of each primary act.

Thus, we can have a *form* whose primary act can be described as: $2^{155}A + 2^{255}A' + 2^{255}B + 2^{255}B' + 2^{255}C + 2^{255}C'$. In any given one of its primary acts, every *form* utilizes all 6 modes of power equally.

The opposite is true of every *generator*. In any given one of its primary acts, every *generator* utilizes a maximum of only 3 modes of power equally, and no 2 of those can be opposites. Thus, we can have a *generator* performing a primary act describable as: $1A + 1A' + 1B'$; but, we can never have a generator performing a primary act describable as: $1A + 1A' + 1B$. That's because A & B are opposites. One can, however, have a generator whose act can be described a 2A *minus* 1B, because opposites can, so to speak, hand over some of their "assets" to one another. They do that in order to insure that—in each and every one of its primary acts—no generator ever activates more than the smallest possible percentage of the power available to it. That smallest possible percentage, of course, is $2^{-256}X$, where X = the maximum amount of power available to each of the 6 modes of power. For an example of what that means, imagine a generator tries to perform a primary act describable as 2A. That's now allowed; and so, it must hand 50% of A over to its opposite B. I'll say no more on that.

D-3. ESOPTRICS' VIEW OF SPACE:

For Esoptrics, space is a function of the forms. For Esoptrics, of course, the word "space" is a misnomer. That's because "space" implies a collection of spaces or, if you prefer, places in space. For Esoptrics, though, space is not a collection of spaces or places. Ultimately (*i.e.:* at a level the lovers of sense images would describe as roughly 10^{-47} cm.), space is a collection of what *primary* acts some form makes possible to a generator. Increase the current OD of a given form, and that form—by means of each of its primary acts—increases the number of primary acts it presents as possible to a generator.

For Esoptrics, particles, atoms, molecules, planets, stars, and all such bodies with mass come about when several forms become logically concentric. That's a result produced whenever a form accelerates above its native OD and, thereby, makes concentric to itself 1 form for each of the OD's thru which it passes. When the form finishes accelerating, it then becomes the leading form for a multitude of forms all having a common center.

Every accelerating form thus causes space to become *stratified*. That means that, for Esoptrics, space is modified by mass in this sense: Each form can—for the sake of the lovers of sense images—be depicted as a larger, *master, composite* cube composed of smaller, *subordinate, non-composite* cubes, and the number of the *subordinate* cubes = $(2D_c)^3$, where D_c = the current OD of the master cube/form. Each of the subordinate cubes represents a particular kind of *primary* act with a generator can perform, and the position of any given subordinate cube within its master cube represents where that *primary* act stands in the logical order of sequence.

Now then, if 2^{155} forms become logically concentric, then—for the sake of the lovers of sense images—we have a master cube inside of, and concentric with, a slightly larger master cube which is also inside of, and concentric with, a slightly larger master cube which is also inside of, and concentric with, a slightly larger master cube, and on and on for 2^{155} steps. Next, each of the master cubes in this concentric mass of 2^{155} master cubes is presenting one or more generators with a definite number of *primary* acts (*i.e.:* subordinate cubes for the lovers of sense images) which one or more generators can perform.

If you've properly pictured this complex of concentric master cubes, then you should be able to perceive this: At the center of this complex, a master cube at OD1 is presenting one or more generators with 8 (*i.e.:* $[2xOD1]^3$) potential *primary* acts. Ah, but! A master cube at OD2 is also presenting the same one or more generators with the same 8 potential *primary* acts as part of its 64 (*i.e.:* $[2xOD2]^3$) subordinate cubes. Ah, but! A master cube at OD3 is also presenting the same one or more generators with the same 8 potential *primary* acts as part of its 216 (*i.e.:* $[2xOD3]^3$) subordinate cubes. Ah, but! Etc., etc., etc., for 2^{155} levels! In other words, for the lovers of sensation and sensation's instruments, we are dealing with only 8 "places in space"; and yet, for Esoptrics and creatures of what we might call "transcendental" logic (*i.e.:* a logic able to transcend sense images), *each* of these 8 "places in space" is 2^{155} "places in space" and all "in the same one place". More correctly, each of these 8 *primary* acts is actually 2^{155} acts. That's because, in one case, the primary act is *in reference to a form at OD1*; in another case, the primary act is *in reference to a form at OD2*; in another case, the primary act is *in reference to a form at OD3*; and on and on for 2^{155} levels. And bear in mind that, according to Esoptrics, there are 6 "places" in the universe at which there are almost 2^{256} master cubes logically concentric with one another, and that means almost 2^{256} generators performing almost 2^{256} primary acts all at what the lovers of sense images would call "a same, one spot only 10^{-47} cm. in length, breadth, and depth". That's Esoptrics' explanation of what "black holes" are.

If you've followed all of the above, you can perhaps now see what Esoptrics means when it says that space, in the vicinity of mass, is *four* rather than merely *three* dimensional. Increase the mass and your closeness to it's center, and you increase the extent to which space is *four* rather than *three* dimensional. That's because, in increments of 10^{-47} cm., as you get closer to the center of a massive body, there's a rise in the number of forms offering you different versions of what the lovers of sense images would call the same, one place in space, and what Esoptrics would call different versions of the same, one primary act. Yes, the primary acts are the same in the terms of which quantity of which mode of power is being activated; but, they are not the same with regard to the same frame of reference.

If you've followed all of the above, you can perhaps now see what Esoptrics means by what Science calls "gravity waves". A form (*i.e.:* a master cube for the lovers of sense images) at $OD2^{256}$ performs a *primary* act whose immediate and instantaneous influence extends—in the terms of lovers of sense images—over an area having a length, width, and depth of roughly 18 trillion light years. By means of that primary act, every generator acting anywhere within that primary act's 5 un-

197

decillion, 840 decillion cubic light years (*i.e.:* $[1.8 \times 10^{12}]^3 = 5.84 \times 10^{39}$ cu. l.y.) is in immediate and instantaneous contact with all the others anywhere within that astronomically vast expanse. Well, for Geometry's lovers of sense imagery, it's an astronomically vast expanse; but, for Esoptrics' lovers of Algebra's transcendental logic, it's merely a single solitary point without the slightest trace of length, width, or depth.[1]

D-4. THE UNRELIABILITY OF THE SENSES:

Let me now close with a bit about trusting the senses. Even in the time of Aristotle, it was commonplace to talk about how misleading sense images are. Even the majority of lesser minds readily noticed how a paddle dipped in water seemed divided in two. That's why, long before the birth of Jesus Christ, men commonly recognized the need to apply critical thinking to the sense images.

Predictably, some went to the other extreme and wound up wholly ignoring sense perception as they concentrated upon *nothing but* hair-splitting debates over the meanings of words. This sort of dialoguing was called "Dialectics". It's a method well illustrated in Plato's dialogues where we find Socrates doing nothing but argue, argue, argue.

Believe it or not, Aristotle (He didn't much care for the method of his former teacher, Plato.) himself complained about those who, in his words, paid too much attention to "Dialectics" and not enough attention to what he called "Science" and "keenly observing what the external world presents to us." Unfortunately, he himself wound up doing so little of what he called Science and so much of what he called "Dialectics", that, when the Catholic philosopher/theologians of the 9th to the 16th century turned to pure Dialectics, they saw themselves as disciples of Aristotle rather than—as was really the case—followers of Plato's love of hair-splitting debates. Granted, they were followers of Aristotle in the sense that they adopted many of his conclusions. So what?! They were by no means followers of the Aristotle who advocated: "keenly observing what the external world presents to us". When it comes to Aristotle, they strained out the gnats and swallowed the camel.

In the face of that aberration, though, one must bear 2 important points in mind: (1) Prior to around 1600 A. D., there were no telescopes, microscopes, thermometers, barometers, and a host of the many other measuring instruments available to us; and (2) prior to around 1450, there were no printing presses able to crank out millions of copies of the latest scientific observations so everybody in every corner of the world could stay up to date on the latest findings. Despite what Aristotle said, prior to 1600, neither he nor anyone else could do much by way of "keenly observing what the external world presents to us." They simply didn't have either the equipment to *observe well* or the equipment to *broadcast widely and*

---

[1] **NOTE OF TUESDAY, SEPTEMBER 14, 2009:** That reminds me of a *Star Trek* episode in which an arch enemy was overcome by enclosing him in a tiny computer chip which, to his view, was an entire universe many billions of light years in diameter. Being a lover of Geometry's sense images, he would never be able to figure out that his astronomically vast universe was actually the opposite of vast; and so, content with his illusion, he would never try to break out and once again menace his would-be victims. Ah, but! What if he were an algebraist named Edward N. Haas???!!!

swiftly the news of what they had observed.

The same lesson is given to us in the passage from Newton to Einstein. Newton ruled the scientific world until telescopes advanced enough to notice that Mercury's orbit was not as fixed as Newton's principles said it should be.

Before truly helpful instrumentation came along, the unaided senses remained so misleading that, for thousands of years, they left humanity wallowing in an incomprehensibly bloody, squalid, and painful deprivation. Science—by improving the yield and reliability of *instrument-assisted* senses—has done much to diminish that misery for a large number of people (It remains to be seen if they've improved the lot of an increased *percentage* of Earth's living occupants.).

Yes, where the senses are not aided by instruments, you are correct to say: "our senses can't be trusted to experience the validity of anything more abstract than bumping our big toe"; but, *instrument assisted* senses can be trusted for far more than that. Still, if Esoptrics is correct, what Esoptrics ultimately proves is this: Without the addition of *pure* reason on the part of a mind able to climb to the highest level of abstract, *a priori* **introspection**, not even Science can achieve the long-sought *complete* step from the world of sensation to the world of:

WHAT REALLY IS.

With all a brother's love:

EDWARD.

ESOPTRICS: LOGIC OF THE MIRROR

Although the greatest divergence of opinion is to be found among philosophers as to the nature of space, yet these opinions are linked together by the degree of reality which they attribute to it; and may be classified under three heads: the theories (1) of those who emphasize this reality, (2) of those who emphasize its ideal character, and (3) of those who preserve the balance between the two. . . . If we make something real of space in itself, we shall be logically led to say that space without bodies is something real, or that absolute space is something real. . . if all bodies are removed, leaving real space behind, this space must be immaterial. It is, however, the function of space to be the location of bodies, and apart from this function we seem to have no conception of it. It is, moreover, clear that the material cannot be located in the immaterial, and so by making space real we deprive it of its only meaning, and render it unintelligible. On the other hand, if we say that space is not real, we shall seem to contradict common sense, to speak in a way which is in disagreement with scientific language, and to deprive the material world of its objectivity.

. . . Democritus, and others . . . taught that space is the vacuum, or a universal receptacle, which is distinct from bodies, but in which they move. Similarly, some Peripatetics taught that space is a kind of immense sphere which surrounds all bodies. Gassendi, in the seventeenth century, reviving the old Atomism, adopted this view of space . . . Locke (1632-1704) also seems to incline to the ultra-realist opinion.

. . . The suggestion that space, the receptacle of bodies, is the vacuum, is untenable, since the bodies are extended, and so their receptacle must be extended also.[1] The vacuum, however, of the Atomists was unextended, and so cannot be space. Even if it were thought of as extended, it would need a further receptacle or vacuum to contain it, and so we should have an unending series of vacua containing one another without ever arriving at an ultimate space. . . an extended thing must be received in something, would apply with equal force to this receptacle itself, and so on to infinity.

. . . Such views of these seem to deify space, and this idea that space is an attribute of the Deity, is found explicitly in the opinion of Newton and Clarke. . . Thus Newton identifies it with God's omnipresence, and Clarke with His immensity. . . . Newton writes: "And these things being rightly dispatched, does it not appear from phenomena that there is a Being incorporeal, living, intelligent, omnipresent, who, in infinite space, as it were in his sensory, sees the things themselves intimately . . . [*Opticks* pp. 344 ff)."
——**R. P. PHILLIPS**: *Modern Thomistic Philosophy*; Vol. I: pgs. 82-96. The Newman Press; Westminster, Maryland; 1956

---

[1] For Esoptrics, space is not the *receptacle* of bodies; as the actuality of a form, it is the *provider* of actual and potential primary acts for one class of ultimate bodies (***viz.:*** the generators). For Esoptrics, both space and the ultimate bodies are *immaterial* and *logically* rather than *physically* extended; and so, both are quite compatible, and neither has any need for an infinite regress. The big problem here is that Fr. Phillips and kind simply cannot even begin to conceive even so little as a single, solitary datum regarding the universe's ultimate, *persistently* real constituents (The lines of tension being our sense images are real but *not persistently* so.). What *really is* lies as far beyond their minds as the ability to picture colored shapes lies beyond one blind from conception. With Esoptrics, all of that comes to an eternal end.

$\mathcal{D}$ocument #12:

A REPLY TO ARISTOTLE ON DISCONTINUOUS TIME & LOCOMOTION:

A. DIVISION IN SPACE VS. IN PRINCIPLE:

Before we can proceed to the main argument, we must lay a little groundwork. For now, then, let us talk about several ways to divide things. In what is about to follow, "space" signifies the extra-mental factor and not merely our intra-mental concept of it. In other words, as used here, "space" never signifies some factor included in one or more of the purely mental states of our minds. If you are one of those convinced there is never anything anywhere but what's included in the purely mental states of your own mind, then let's not bother each other.

Things can be divided either *in space* or *in principle*. Between those two kinds of division, the difference is exceedingly radical. How is that?

Consider first the division **in space**. It applies to extra-mental, physically extended items, which is to say items we commonly class either as space or extra-mental *objects*. The latter have 2 sets of factors we can observe either with our senses alone or with the aid of telescopes, microscopes, and the like. Those 2 sets of observable factors are: extremities and the material traits they enclose.

Such *objects* can be divided in space either in fact or in fancy or both. If you prefer, say they can be spatially divided either in thought or in reality or both. To put it yet a third way, say they can be divided in space either intra-mentally or extra-mentally or both. Fourthly, they're divisible mentally and/or factually in space.

By way of example, consider our home planet. Though some future generation may eventually gain such power, no one on Earth yet has the ability to split it in *fact* into two halves in *space*. So far, we can only *imagine* the Earth spatially split into two halves, which is to say that only intra-mentally in fancy and thought can we currently divide it in space. Conversely, given an orange, one can either be content to *imagine* it spatially sliced into two halves, four quarters, etc., or one can take a knife in hand and actually *slice* it into some chosen number of pieces which can then be spatially separated from one another in fact.

How is it that we are mentally and/or factually able to divide extra-mental objects in space? It's because they themselves are already factually divided in space in the sense that they have what our senses observe as material parts outside of one another in space. In referring to them as *material* parts, I advert to the fact that each part presents our senses with material-like traits.

What shall we say of space itself? Can it—in fact and/or fancy—be divided in space? As far as our naked senses can observe, space is merely the absence of any source of sense images. It's a lacuna in the midst of two or more extra-mental physical objects. As an extra-mental object, space is by no means *physical*, which is to say it has no observable extremities enclosing material-like traits. As a result, one cannot *in fact* split space into two or more parts in space. For, what would lie between them as you moved them away from one another? One can, however, *in fancy* divide space into two or more spaces outside of one another in space. To do that, though, one must use extra-mental, physical objects or their intra-mental impressions to mark off the divided spaces. Space, then, is a kind of oddball target of division. It is the *medium* of division in space; but, since it is merely the *absence* of all sources of sense images rather than itself such a source, we cannot—by using space alone—divide space either mentally or factually.

Most assuredly, space is in fact divided in the sense that it has parts outside of one another. Those parts, though, are outside of one another in *space*, and that leaves us unable to divide them *mentally* without the aid of some factor very different from space itself. It's a mutual benefit situation: We cannot either mentally or factually divide material objects into their parts without the aid of space, and we cannot mentally (Factually is out of the question.) divide space into its parts without the aid of material objects.

In sum, then, what is divid*ed* in space is whatever strikes our senses as an extra-mental, physically extended factor—whether it be a material object or space itself. It is then divis*ible* by us, if we are informed enough to know how to divide it in space whether mentally or factually or both.

. Consider next the division **in principle**. Division in principle means we can mentally break targets down into more than one *principle*. For example, we can speak of some or all material objects as having both potentiality and actuality, as having both matter, a leading form, and generators. We can also speak of them as having weight, shape, and color. In each of those cases, one is *mentally* dividing extra-mental factors in *principle*. In none of those cases are we—whether in fact or in fancy—*truly* dividing them in *space*. After all, whether in fact or in fancy, we cannot truly introduce a spatial division between actuality versus potentiality or matter versus form or shape versus its owner or generators versus the lines of force they generate. We cannot do so, because our minds cannot produce an extra-mental object which is *solely* a duplicate of the internal characteristics of actuality itself or *solely* a duplicate of forms themselves or of generators themselves or of shapes themselves. Our inability to produce such extra-mental objects stems from the fact that we cannot extra-mentally reproduce what is no more than a duplicate of what—though it may or may not itself be some kind of reality—is, for us, a pure abstraction known only or mostly in the terms of its relations.

So then, intra-mental division in principle is a procedure we perform upon things which are—at least for us—abstractions, which is to say things known to us solely or mostly in the terms of their relations. We can perform division in principle on any target of thought as long as one crucial condition is fulfilled in that target. That condition is this:

MORE THAN ONE PRINCIPLE IS CONJOINED
WITHIN THE BOUNDARIES OF THAT TARGET.

We have said that we cannot—whether factually or mentally—*truly* introduce an extra-mental division in space into that which is divided only in principle, which is to say that which is, for us, purely an abstraction. That does not rule out *figuratively* introducing an extra-mental division in space into abstractions. To produce such a *figurative* division in space, all we need is one or more extra-mental, physical objects which—though they fail to replicate the *internal characteristics* of the targeted abstraction—do diagram its *relations*. By way of example, most city street maps are not photographs of the city or its streets; instead, they merely use a diagram to express figuratively the way the various streets are related to one another. To be sure, neither the city nor its streets are merely abstractions; and so, the example limps badly. Do you, nevertheless, catch my drift?

Now then, it's easy enough to realize that *one* given thing has relations with some *other* given things. It's not so easy to realize that, if a given thing is composed of two or more principles, it has relations with *its own self*. More correctly, the *principles* united in it can have relations to one another. Therefore, given an abstraction in which two or more abstract principles are united, it is possible to bring that abstraction into focus with a drawing which does no more than serve as a diagram *figuratively* expressing the relations between the two or more abstract principles united in it. In turn, that gives us a drawing which we can then intra-mentally and/or extra-mentally divide in space. The instant we execute such spatial divisions upon the drawing, we then *figuratively* divide in space—whether mentally or factually or both—an abstract factor which cannot *truly* be divided by us save intra-mentally in principle.

That, of course, is exactly what we have seen done many times in the course of this book. In drawings such as drawing #7 on page 32, we have seen diagrams figuratively expressing forms and generators in the terms of the relations which hold between their actualities, their 6 kinds of potentiality, and the 2^{256} levels of intensity which each of those 6 modes of power can employ. A man conceived without eyes cannot have the slightest hint of what its like to experience light. By the same token, an intelligent being without access to infinity cannot have the slightest hint of what its like to experience the internal characteristics of the primary act of a form or a generator. For our drawings, it's an irrelevant fact. For, our drawings do not make even the slightest attempt to present their viewers with an experience of the internal characteristics of some primary act. All they propose to present is an experience of how to diagram certain relations which prevail between the seven (or 7×2^{256}, if you prefer) principles found in every primary act. At the same time, neither do they propose to present us with an experience of those relations as they absolutely are to God. All they propose to do is to present us with an experience of which diagrams can figuratively and most efficiently express those relations to intelligent beings who *directly* experience nothing but sense images.

In sum, then, what is divid<u>ed</u> in principle is whatever encompasses within itself 2 or more distinct principles. It is then divis<u>ible</u> by us if we have enough knowledge of those principles to express figuratively and diagrammatically how those principles are related to one another.

How does time fit into this picture? Time is an abstraction in which 3 principles are conjoined—namely: past, present, and future. Time is thus divided and divisible by us in principle. At least, that's the case as long as "time" signifies a *length* of time, which is to say a time span with some duration. *Durationless instants* of time are absolutely undivided and indivisible whether in space or in principle. They can never, by themselves, add up to a time span having a duration.

Have you grasped all I've said so far in this section? If so, you can perhaps perceive this: Nothing is *absolutely* indivisible, unless—both mentally and factually—it is neither divided nor divisible either in space or in principle.

What, now, shall we say of a point? For Aristotle and perhaps every one else in recorded history (myself included), "point" has always signified what has no trace of *physical* extension, which is to say what has no trace of length, width, or depth. As such, every point is *absolutely* indivisible in *space*.

But, is every point *absolutely* indivisible in *principle*? Prior to Esoptrics, the answer was yes from Aristotle and every one else known to history. Why?! It's because, prior to Esoptrics, no one had the slightest idea of what *two or more principles* could be conjoined in a point, and that necessarily left them with no awareness of how a point could be divisible *in principle*. To be specific about it, prior to Esoptrics, no one had the slightest bit of knowledge regarding the 6 modes of power or their 2^{256} levels of intensity or their ways of relating to one another. Indeed, they had not even the slightest suspicion or shadow of an inkling that there could be such factors. Such factors were as far above their minds as Relativity was above the mind of a Neanderthal fetus. Such factors were *so* far above their minds, it strongly suggests they—at least at some totally unconscious, instinctive level essential to our inheritance—didn't *want*, and were *adamantly opposed to having*, knowledge of such factors.

There are, therefore, 2 radically different kinds of points. One kind is—*purely and simply so!*—absolutely indivisible, because it is absolutely indivisible both in space and in principle. The other kind is only *in a manner of speaking* absolutely indivisible, because, while it is absolutely indivisible in space, it is by no means indivisible in principle. Hereafter, absolutely indivisible shall denote only the former.

Above all, what is mainly to be kept in mind here is this: Prior to Esoptrics, there was nowhere any knowledge of the second kind of point, and the first kind of point was the only kind of point known either to Aristotle or anyone else known to history. Once that is realized, Aristotle can be refuted as easily as falling off a log.

B. ARISTOTLE'S ARGUMENT:

Aristotle writes:

To material masses, to time, and to motion, the same line of reasoning applies with equal force. It goes like this: Either each of them is composed of indivisibles and is divisible into indivisibles, or none of them are. That may be clarified so: If a material mass is composed of indi-

visibles, then a motion over that mass must also be composed of corresponding indivisible motions. For example, imagine a mass called ABΓ composed of the indivisibles A, B, and Γ. Next, Ω moves across it with a motion called ΔEZ. Each corresponding part of ΔEZ across ABΓ is indivisible. Why?! Where there is motion, there must be something moving, and where something is moving there must be motion; and so, the motion, too, will be composed of indivisibles. We must say, then, that Ω crosses A as its motion is Δ, crosses B as its motion is E, and crosses Γ as its motion is Z. Now then, what's in motion from one place to another cannot, while moving, be both moving toward, and, simultaneously, finished moving toward, its destination. For example, suppose a man is walking to Thebes. He cannot simultaneously be both walking to Thebes and finished with his walk to Thebes; and yet, that's what we see if Ω crosses the partless section A by means of the motion Δ . . . . it is passing thru and has completed its passage *at the same instant*. Thus, that which is walking shall—at the instant when it is walking—have completed its motion at the place to which it is in motion. . . . it has completed its passage thru A without passing thru it. It shall thus be possible for a thing to have completed a walk without ever walking. For, on the current assumption, it has completed a walk over a particular distance without walking over that distance. Therefore, since everything must be either at rest or in motion (and yet, Ω is at rest in each of the sections A, B, and Γ), it follows that a thing is—*simultaneously!*—both continuously at rest and continuously in motion. For, we saw that Ω is both moving across the whole of ABΓ and resting in every part of it and, therefore, resting in the whole of it.
——*Physics*; Book VI: Chap. 1; 231^b:18-30 & 232^a:12-14. My rather free translation - ENH. Compare page 313 of vol. 8 of ***Great Books Of The Western World***, published by William Benton, Chicago, 1952.

What a powerful argument indeed! Alas for Aristotle! It's potent only as long as you fail to notice he's neglected to mention at least *half* the moments involved. He has, you see, listed only the three timeless moments of duration and not a one of the three or even four durationless moments of time which belong in the picture. That's the *gist* of my reply. Let's now spell it out in detail.

Aristotle refers us to some particular mass to which he gives the name "ABΓ," because it is composed of three indivisible parts—one of which is called "A," the second of which is called "B," and third of which is called "Γ". Someone or something is said to be moving across that mass. To this traveler, he gives the name "Ω".

If any part of that is confusing to you, turn to figure #2 on page 6. Notice the three points: 0, A_1 and A_2. You can easily rename those points A, B, and Γ. The line running from 0 to A_2 through A_1 then becomes the mass ABΓ. The traveler, Ω, then becomes someone or something moving along the line $0A_2$ by first jumping from 0 (renamed A) to A_1 (renamed B) and then from A_1 to A_2 (renamed Γ). In this book's terminology, Ω represents a generator performing three different acts in succession.

ESOPTRICS: LOGIC OF THE MIRROR

As Ω moves across ABΓ, Aristotle says there is a moment in which Ω crossed A. It is followed by a second moment in which Ω crossed B. That moment, in turn, is followed by a third moment in which Ω crossed Γ. The first moment, Aristotle calls "Δ". The second, he calls "E". To the third, he gives the name "Z". In short, the only moments mentioned by Aristotle are Δ, E, and Z.

I say: Realize the difference between the *two* kinds of points and indivisibles, and you rather quickly see there are four *other* moments which need to be described. Let us give them the names "C", "F", "H", and "J".

Before we describe C, F, H, and J, let's clarify another detail. Aristotle *first* describes Ω as *crossing* something in the moments, Δ, E, and Z, and, thereby, makes them moments of *locomotion*. In the end, though, he winds up calling them moments of *rest*. I see no need to repeat his explanation of why he does so. What's important for now is that his final position was this: During the moment Δ, Ω is nowhere but at *rest* in section A; during the moment E, Ω is nowhere but at *rest* in section B; and during the moment Z, Ω is nowhere but at *rest* in section Γ.

From the standpoint of the moments Δ, E, and Z, then, there is no "locomotion", and Ω is, rather, at "rest" in each of the sections A, B, and Γ in succession. For Esoptrics, however, it cannot be said that, while at each of those sections, Ω is doing nothing but rest. After all, in each given section, Ω is performing the *primary act* unique to the given section, and one must not imagine that's to any extent akin to the visual image of an inert hardness (such as a rock) standing *dead still* in one spot on the ground. The far more accurate description of the moments Δ, E, and Z would be this: During Δ, Ω is performing the *primary act* which constitutes the location A; during E, Ω is performing the *primary act* constituting B; and during Z, Ω is performing the *primary act* constituting Γ.

We can now explain the moments C, F, H, and J. During C, Ω has just finished performing the primary act constituting some immediately preceding location and is about to perform the primary act constituting the location A. During F, Ω has just finished performing the primary act constituting the location A and is about to perform the primary act constituting the location B. During H, Ω has just finished performing the primary act constituting the location B and is about to perform the primary act constituting the location Γ. During J, Ω has just finished performing the primary act constituting the location Γ and is about to perform the primary act constituting some next location.

C, F, H, and J are the moments of *local movement*; Δ, E, and Z are <u>not</u>. The reason for that distinction should be quite manifest: During each of the moments Δ, E, and Z, there is, for Geometry, no change of place in space and, for Esoptrics, no change from act to act. Oppositely, during each of the moments C, F, H, and J, Ω is in no given location (no primary act) whatsoever; it is *between* locations (primary acts) and, consequently, is truly in the process of transiting from one location (primary act) to another. Esoptrics expresses that difference by saying that Δ, E, and Z are moments of the *primary* act most properly called *existence* and *motion*; whereas, C, F, H, and J are moments of the *accidental* activity Geometry calls *local movement* or *locomotion* and Esoptrics calls *transition from primary act to primary act*.

The moments Δ, E, and Z are *enduring* ones. They are moments which have a duration which can be relatively expressed as a definite number of alphakronons. That's but another way of saying each of those three moments lasts for as long as Ω is engaged in performing the same, one, specific primary act.

C, F, H, and J are *non*-enduring moments. They are moments without duration. They are durationless moments. That is but another way of saying that there is no primary act in the durationless course of either C, F, H, or J. Each of the four merely distinguishes between: (1) when one primary act ended, and (2) when another one commenced.

Δ, E, and Z are moments of changelessness. In the *course* of each of them, there is no change of location (primary act) and, as we've seen, no change either in the quantity of potency being actualized or in the co-ordinates defining that actualized potency. Since time is commonly defined as the measure of change (and it is certainly such for Esoptrics), Δ, E, and Z cannot be moments of time. They are, rather, enduring moments of timeless changelessness. More succinctly put, they are timeless moments of duration.

C, F, H, and J are, by definition, moments of change. Since time is the measure of change, C, F, H, and J are moments of time. Since they are without duration, they are durationless moments of time. Because they are durationless moments of time, C, F, H, and J are absolutely indivisible points. Whether mentally or factually, they cannot be divided either in space or in principle.

Because they pertain to an enduring primary act, Δ, E, and Z are by no means absolutely indivisible points. To be sure, they are timeless moments of duration which—whether mentally or factually—cannot be divided in space; but, they can be mentally divided in principle. That's because they can be mentally divided into a definite number of alphakronons, and what's occurring during each can be mentally divided into as many as 7×2^{256} principles.

Aristotle writes:

> Let the terms "continuous", "in contact", and "in succession" be taken as we defined them earlier, namely: Things are continuous if they have the same extremities; they are in contact if their extremities are juxtaposed; and they are in succession if there is nothing of their own kind intermediate between them.
> ——*Physics*; Book VI: Chap. 1; 231[a], 20-25. My rather free translation - ENH. Compare page 312 of vol. 8 of **Great Books Of The Western World** as published by William Benton, Chicago,1952.

For the moment, we are only concerned with how Aristotle defines "in succession." Look closely at the above quote, and you see he tells us things are in succession "if there is nothing of their own kind intermediate between them." Remember that, and remember it well.

C, F, H, and J are durationless moments of time, and the only moments "intermediate between them" are the three timeless moments of duration designated Δ, E, and Z. I need hardly point out that durationless moments of time and timeless moments of duration are very much *opposites* in *kind*. C, F, H, and J, then, are *in succession*, because, in Aristotle's own words, "there is **NOTHING** *of their **OWN** kind* **INTERMEDIATE** between them.*" By the same token, Δ, E, and Z are *in succession*, because, in Aristotle's own words, "there is **NOTHING** *of their **OWN** kind* **INTERMEDIATE** between them.*"

The sum of it all so far, then, is this: Aristotle is wrong when he says the time span is ΔEZ and includes only the indivisible instants of time he calls Δ, E, and Z. To be correct he must say the time span is, and includes, C, F, H, and J. To be even more precise, he must say CΔFEHZJ includes the durationless moments of time called C, F, H, and J alternating with the timeless moments of duration called Δ, E, and Z. He would then also have to point out that, though C, F, H, and J are each *absolutely* indivisible points, Δ, E, and Z are not.

C. REFUTATION OF ARISTOTLE'S ARGUMENT:

We can now return to Aristotle's argument and refute it. In the lengthy quote given at the beginning of section B, he related a line of thought quite similar *at first sight* to what Esoptrics proposes. He then sought to explode such thinking by convicting it of producing a manifestly self-contradictory and absurd conclusion. He sought to show that, from such thinking, "it follows that a thing is—*simultaneously!*—uninterruptedly at rest and uninterruptedly in motion."

If you've followed what I've said in the preceding section, B, you now perceive that Aristotle's conclusion by no means pertains to Esoptrics. For, what Esoptrics says is this: Given the same one time span, a thing is:

(1) CONSISTENTLY PERFORMING A DIFFERENT ACT

with regard to each moment in an unbroken succession of timeless moments of duration,

AND

(2) CONSISTENTLY CHANGING FROM ONE DIFFERENT ACT TO ANOTHER

with regard to each moment in an unbroken succession of durationless moments of time.

Most manifestly, in such a proposal, one cannot, as Aristotle would have it, find the grossly absurd, self-contradictory nonsense of saying that—throughout the course of the same one time span and *simultaneously*—a thing can be uninterruptedly at rest and in motion. For, the so-called "at rest" is with regard to one series of successive moments, and the so-called "motion" is with regard to a second, very different series of successive moments. It's merely a question of realizing that any given time span involves *two* concurrent streams of *two* radically different kinds of successive moments rather than *one*. Of course, the ability to arrive at such a realization depends upon a prior ability to arrive at the realization that there are *two* kinds of *points* rather than *one* and *two* kinds of *indivisibles* rather than *one*. But, whether one can arrive at such realizations or not, the bottom line remains untouched, and that bottom line is this: Aristotle's argument obviously has nothing to do with Esoptrics' take on discontinuous time and locomotion.

D. A TIME SPAN VS. AN INSTANT OF TIME:

Suspecting such an answer will by no means satisfy *all*, I will approach the issue from a different angle. To that end, let's first note two very different ways to use the word "time".

On the one hand, "time" can refer to a particular time span (*i.e.: length* of time) falling in between two arbitrarily chosen *instants* of time—one of which serves as the beginning of the time span; the other of which serves as the end. For example, it could refer to a particular day, a particular year, a particular century, a particular epoch of creation, a particular season, or even the history of some particular church, nation, ideology or species. In each such case, one would be denoting a kind of time which has a **duration** lasting <u>from</u> some at least hypothetical *instant* of time—an instant serving as the beginning—<u>to</u> some other at least hypothetical *instant* of time serving as the end.

As we've just seen, then, "time" can signify an *instant* of time for all. For example, it could designate either the instant at which a particular day, year, century or epoch began or the instant at which they ended. In each such case, one would be pinpointing a kind of time which has **no duration** and which is absolutely indivisible in the purest sense (*i.e.:* absolutely indivisible in space and in principle).

Now, take that distinction between *lengths* of time and *instants* of time, and put it on hold while we talk about the word "continuous". "Continuous" does not *always* imply what's so joined as to rule out *all* interruptions and intervening stops. By way of illustration, in defining the word "continued", **Webster's Deluxe Unabridged Dictionary** (Second Edition) speaks of a "continuous voyage" as one *considered* such because—despite many intermediate stops—its *final* destination was one throughout the trip. By the same token, falling back to some extent upon Aristotle's own example of the man enroute to Thebes, we can say that a man *limping* all the way to Thebes would be described as *continuously* limping along the way; and yet, properly speaking, the limp occurs only as he drags the bad leg up to the good one—at which point, he would pause very briefly and step off rather normally with the good leg.

Speaking, then, of a man limping toward Thebes for, say, twenty-four hours straight, it would be commonplace to say he was—within the same, one time span— continuously moving toward Thebes in a stop and go manner. For Esoptrics, *that* is how all changes through time and space are at least *effectively* described.

It's far from likely Aristotle was objecting to such a notion. Were the opposite true, Aristotle's argument would be worthless. After all, it may be rather *sloppy* to speak of a *methodically* interrupted motion toward one's chief goal as *continuous*; still, it's far from *self-contradictory* or *patently absurd*.

No, Aristotle's is not an argument against local movements in which pauses alternate with local movements. Rather, his argument is against local movements in which every indivisible instant thereof claims to be both a pause and a local movement simultaneously. To illustrate how such self-contradictory claims come about in Aristotle's view, we can rightly rephrase his argument so: "If you say every timespan and/or motion is a chain of indivisibles, you necessarily reduce every timespan and/or motion to a chain of *absolutely* indivisible units which, as such, are *durationless*. Therefore, whenever you say every timespan is a chain of indivisible instants of time, you reduce every time span to *nothing but durationless* instants. If you then add that a certain traveler is in motion at every durationless instant of a particular timespan, you engage in self-contradictory gibberish. For, every timespan necessarily has a duration; and yet, no number of duration*less* instants can ever add up to a duration. The same is true of every given instance of locomotion: Every such necessarily has a duration; and yet, no number of durationless instants of motion can add up to a duration and, consequently, can never add up to a motion. Necessarily, where there is no motion, there is only rest; and so, if you reduce every timespan and motion to indivisible units, you say every instance of locomotion is *not* locomotion and is rest instead. But, rest, too, is not rest unless it has a duration, and no number of durationless instants of rest can add up to a duration. In short, if you say every timespan and motion is a chain of durationless units only, you say every instance of locomotion is simultaneously locomotion but not locomotion and rest but not rest. That being so, one cannot sanely say either: (1) that every timespan is a chain of nothing but durationless instants of time, or: (2) that every locomotion is a chain of nothing but durationless motions."

Such an argument has nothing to do with what Esoptrics says about time and local movement. After all, the above argument is based upon a *false* assumption—namely: Whenever you say every timespan and motion is a chain of absolutely indivisible units, you invariably reduce every timespan and motion to *nothing but* **one** kind of indivisible unit—namely: a *durationless* one. To that, Esoptrics gives the lie quite thoroughly. For, though Esoptrics says every timespan and motion is a chain of durationless units, it *also* says every timespan and motion is *also* a concurrent chain of timeless moments of duration. In other words, Esoptrics says every timespan and motion is two concurrent chains of *two* kinds of moments, points and indivisibles. Only *one* of those two kinds involves *durationless* instants; the other of those two involves the *opposite* of durationless instants. In the face of Esoptrics, then, it is patently false to say: Whenever you say every timespan and motion is a chain of absolutely indivisible units, you invariably reduce every timespan and motion to *nothing but* **one** kind of indivisible unit—namely: a *durationless* one.

Here again, then, we are faced with what should now be a familiar problem: Aristotle could not conceive of *two* kinds of points and indivisibles. He could conceive only of the kind which is absolutely indivisible both in space and in principle; and so, for him, to say every timespan and motion is a chain of indivisibles is necessarily to say every timespan and motion is a chain of only one kind of indivisible. Let him say it all he wants. In the face of Esoptrics, his assertion is patently false to all save to those who, like Aristotle, deny there is any such thing as an indivisible point actually divided into several principles without those principles being divided in space. Why would anyone make such a denial?

In post Esoptrics days, what could the answer be save ignorance too gross to be excusable? That could not be true of Aristotle. For, in his day, Esoptrics was not known. There was no knowledge of the 6 modes of power or their 2^{256} levels of intensity or their ways of relating to one another. Without that knowledge, there was no way Aristotle could have had the slightest inkling of what *multiple* principles could be conjoined in a point and, consequently, could not have had the slightest inkling of how a point could be divided and divisible in *principle* regardless of how indivisible it might be in *space*.

On the other hand, there is this fact: Any comparison of my writings with Aristotle's shall always and conclusively prove his mental acuity many times greater than mine. How, then, was it possible for the far superior genius of Aristotle (not to mention that of perhaps millions of others, St. Thomas Aquinas in particular) to miss throughout his life what I, as young as 20, missed _not_?[1] Maybe it's just another example of the astonishing proficiency with which primitive imagery imperceptibly draws virtually every human being into unconsciously swallowing assumptions of the most ludicrous kind. If so, how does primitive imagery have such startling power?

It's not a question to be considered here. Here, the all-important issue is this: Once you assume that every effort to reduce a timespan or motion to indivisibles *must* result in a *single* stream of a *single kind* of moment, only the densest of morons could fail to prove that, whenever you try to reduce time, space, matter, and locomotion to indivisibles, it leads to conclusions too absurd to be even remotely tenable. Nevertheless, it _is_ merely a *blind assumption*, and, for that reason, Aristotle and company have, in the final analysis, proven naught save this: What can *only* be explained in the terms of *two* concurrent streams of *two* kinds of indivisibles is utterly incapable of being explained in the terms of *one*. Personally, I don't consider that a remarkable accomplishment.[2]

---

[1] **NOTE OF OCT. 13, 2009:** Earlier today at a reunion of old friends, one of my buddies from grammar school and high school listened to me talking about the two kinds of points and said: "That's the same as what you used to tell us in the eighth grade, and I still can't understand any of it." Was I, then, as young as *twelve* when I missed not what far greater geniuses did?

[2] In my view, no time span is absolutely continuous, unless there is no limit to the smallness of the smallest segment of it, which is to say it must *actually* include a segment $10^{-\infty}$ sec. in duration. Such a segment, though, would be durationless and absolutely indivisible, and, therefore, no number of them in succession could ever produce a time span having a duration. By the same token, no area of space or of a spatial object is absolutely continuous, unless there is no limit to the smallness of the smallest segment of it, which is to say it must *actually* include a segment $10^{-\infty}$ cm. in diameter. Such a segment, though, would be devoid of physical extension and would be absolutely indivisible, and, therefore, no number of such areas could ever produce a physically extended area. By the same token, no instance of locomotion is absolutely continuous, unless there is no limit to the smallness of the smallest segment of it, which is to say it must *actually* include a segment $10^{-\infty}$ cm. in length. Such a segment, though, would be absolutely indivisible and devoid of motion, and, therefore, no number of such segments in succession could ever produce a motion of any length whatsoever. It must be, then, that neither time nor space nor material objects nor locomotion can possibly be infinitely di-

E. A VISIBLE OBJECT AT REST IS NOT OUTSIDE OF TIME:

Let me not close without tending to a minor point. The very astute reader might object: "Suppose Ω arrives at Thebes and sits down in a certain spot. In that case, Ω is steadily at rest in one place for as long as he sits there. But, if, as you say, every durationless moment of time involves transition from place to place, then Ω is either: (1) no longer involved in time as long as he remains seated, or (2) simultaneously resting in one place and moving from place to place. Choose whichever alternative you wish; you're left with an absurdity. Seeing your thinking produces such ridiculousness, it is untenable."

It is merely a matter of reminding my reader of what has already been rather well explained in earlier pages: Motion through time pertains either to forms or to generators. Forms change from act to act without in any way changing from place to place. Only of a carrying generator is being seated in place in Thebes a *plausible* issue. It's hardly a *viable* one, because it's easily dismissed so: When generators change from primary act to primary act (and thereby move from place to place in the view of sense perception), their motions are so small (*i.e.* 10^{-47} cm. in the view of sense perception), billions could occur per picosecond without producing a movement detectable by any *human* observer watching a *visible* object resting in one spot in Thebes. The objection, therefore, is without merit, and my refutation of Aristotle stands.

✝✡✝✡✝✡✝✡✝✡✝✡✝✡✝✡✝✡✝✡✝✡✝

. . . . mathematics cannot really handle infinite numbers . . . .
——**STEPHEN HAWKING:** *A Brief History Of Time*, as found on page 49 of the paperback, updated and expanded, tenth anniversary edition published by Bantam Books, New York, 1998. © 1988, 1996 by Stephen Hawking. [I've had my fill of scientists and mathematicians saying the very opposite. How refreshing it is, then, to hear one of Prof. Hawking's stature make such a confession!]

---

visible and, consequently, cannot possibly be absolutely continuous. It must be, then, that—at some most fundamental level—time, space, material objects, and locomotion reduce to some factor which is no way divisible in space but very divisible in principle. For me, those are among the most immediately and glaringly obvious inferences given to the human mind. Consequently, as I see it, all mentally competent thinkers at all times everywhere are always confronted with an exceedingly powerful incentive to look for that "factor which is no way divisible in space but very divisible in principle". For all of that, history suggests to me that, in the whole of pre-Esoptrics times, not a single solitary human being ever did so little as acknowledge the presence of such an incentive, let alone make any effort whatsoever to respond to it. In my view, that fact suggests a well-hidden, ulterior motive so gross, it may, on Judgment Day, stand as a terrible blot upon every pre-Esoptrics thinker's claim to intellectual integrity.

APPENDIX A:

CHART OF CRUCIAL VALUES

UNIT	SIGN	DESCRIPTION
Alphakronon	K	$7.201789375 \times 10^{-96}$ seconds.
Alphatopon	Φ	$7.34683969 \times 10^{-47}$ cm. = 2^{-129} times the atomic dia..
Atomic Dia.	none	Diameter of the hydrogen atom with the electron in its outermost orbit = 5×10^{-8} = 1/20,000,000 = .00 000 005 cm.
Centimeter	cm.	$1.361129468 \times 10^{46}$ Φ
Dia. Of The Form Of The Universe	D_u	1.7014118×10^{31} cm. = 1.79844×10^{13} light years = 2^{128} times the atomic diameter = 2^{257} Φ
Duration Of The Ninth Ep-och	none	2^{385} K = 5.67532×10^{20} sec. = 1.79844×10^{13} Earth years.
Earth Day Of 360°	none	23 hrs., 56', 4.09" = 86,164.09 sec.
Earth Day Of 360°	none	1.19636×10^{100} K.
Light Speed In Cm./Second	none	29,979,245,800 cm./sec. ± 40,000 cm./sec.
Light Speed In K/Φ	none	2^{128} K/Φ
Light Speed In Φ/Sec.	none	$4.080563489 \times 10^{56}$ Φ/sec.
Light Year	none	9.46051×10^{17} cm. = 5.878477×10^{12} mi. per 31,556,925.9747 sec.
Second	sec.	$1.388543802 \times 10^{95}$ K

213

UNIT	SIGN	DESCRIPTION
Tropical Year	none	31,556,925.9747 sec.
Tropical Year	none	4.38158×10^{102} K

APPENDIX B:

POWERS OF 2

$2^0 =$ 1
$2^1 =$ 2
$2^2 =$ 4
$2^3 =$ 8
$2^4 =$ 16
$2^5 =$ 32
$2^6 =$ 64
$2^7 =$ 128
$2^8 =$ 256
$2^9 =$ 512
$2^{10} =$ 1,024
$2^{11} =$ 2,048
$2^{12} =$ 4,096
$2^{13} =$ 8,192
$2^{14} =$ 16,384
$2^{15} =$ 32,768
$2^{16} =$ 65,536
$2^{17} =$ 131,072
$2^{18} =$ 262,144
$2^{19} =$ 524,288
$2^{20} =$ 1,048,576
$2^{21} =$ 2,097,152
$2^{22} =$ 4,194,304
$2^{23} =$ 8,388,608
$2^{24} =$ 16,777,216
$2^{25} =$ 33,554,432
$2^{26} =$ 67,108,864
$2^{27} =$ 134,217,728
$2^{28} =$ 268,435,456
$2^{29} =$ 536,870,912
$2^{30} =$ 1,073,741,824
$2^{31} =$ 2,147,483,648
$2^{32} =$ 4,294,967,296
$2^{33} =$ 8,589,934,592
$2^{34} =$ 17,179,869,184
$2^{35} =$ 34,359,738,368
$2^{36} =$ 68,719,476,736
$2^{37} =$ 137,438,953,472
$2^{38} =$ 274,877,906,944

$2^{39} =$ 549,755,813,888
$2^{40} =$ 1,099,511,627,776
$2^{41} =$ 2,199,023,255,552
$2^{42} =$ 4,398,046,511,104
$2^{43} =$ 8,796,093,022,208
$2^{44} =$ 17,592,186,044,416
$2^{45} =$ 35,184,372,088,832
$2^{46} =$ 70,368,744,177,664
$2^{47} =$ 140,737,488,355,328
$2^{48} =$ 281,474,976,710,656
$2^{49} =$ 562,949,953,421,312
$2^{50} =$ 1,125,899,906,842,624
$2^{51} =$ 2,251,799,813,685,248
$2^{52} =$ 4,503,599,627,370,496
$2^{53} =$ 9,007,199,254,740,992
$2^{54} =$ 18,014,398,509,481,984
$2^{55} =$ 36,028,797,018,963,968
$2^{56} =$ 72,057,594,037,927,936
$2^{57} =$ 144,115,188,075,855,872
$2^{58} =$ 288,230,376,151,711,744
$2^{59} =$ 576,460,752,303,423,488
$2^{60} =$ 1,152,921,504,606,846,976
$2^{61} =$ 2,305,843,009,213,693,952
$2^{62} =$ 4,611,686,018,427,387,904
$2^{63} =$ 9,223,372,036,854,775,808
$2^{64} =$ 18,446,744,073,709,551,616 $= 1.8447 \times 10^{19}$
$2^{65} =$ 36,893,488,147,419,103,232
$2^{66} =$ 73,786,976,294,838,201,464
$2^{67} =$ 147,573,952,589,676,412,928
$2^{68} =$ 295,147,905,179,352,825,856
$2^{69} =$ 590,295,810,358,705,651,712
$2^{70} =$ 1,180,591,620,717,411,303,424
$2^{71} =$ 2,361,183,241,434,822,606,848
$2^{72} =$ 4,722,366,482,869,645,213,696
$2^{73} =$ 9,444,732,965,739,290,427,392
$2^{74} =$ 18,889,465,931,478,580,854,784
$2^{75} =$ 37,778,931,862,957,161,709,568
$2^{76} =$ 75,557,863,725,914,323,419,136
$2^{77} =$ 151,115,727,451,828,646,838,272
$2^{78} =$ 302,231,454,903,657,293,676,544
$2^{79} =$ 604,462,909,807,314,587,353,088
$2^{80} =$ 1,208,925,819,614,629,174,706,176
$2^{81} =$ 2,417,851,639,229,258,349,412,352
$2^{82} =$ 4,835,703,278,458,516,698,824,704
$2^{83} =$ 9,671,406,556,917,033,397,649,408
$2^{84} =$ 19,342,813,113,834,066,795,298,816
$2^{85} =$ 38,685,626,227,668,133,590,597,632
$2^{86} =$ 77,371,252,455,336,267,181,195,264
$2^{87} =$ 154,742,504,910,672,534,362,390,528

2^{88} = 309,485,009,821,345,068,724,781,056
2^{89} = 618,970,019,642,690,137,449,562,112
2^{90} = 1,237,940,039,285,380,274,899,124,224
2^{91} = 2,475,880,078,570,760,549,798,248,448
2^{92} = 4,951,760,157,141,521,099,596,496,896
2^{93} = 9,903,520,314,283,042,199,192,993,792
2^{94} = 19,807,040,628,566,084,398,385,987,584
2^{95} = 39,614,081,257,132,168,796,771,975,168
2^{96} = 79,228,162,514,264,337,593,543,950,336 = 7.922816×10^{28}
2^{97} = 158,456,325,028,528,675,187,087,900,672
2^{98} = 316,912,650,057,057,350,374,175,801,344
2^{99} = 633,825,300,114,114,700,748,351,602,688
2^{100} = 1,267,650,600,228,229,401,496,703,205,376
2^{101} = 2,535,301,200,456,458,802,993,406,410,752
2^{102} = 5,070,602,400,912,917,605,986,812,821,504
2^{103} = 10,141,204,801,825,835,211,973,625,643,008
2^{104} = 20,282,409,603,651,670,423,947,251,286,016
2^{105} = 40,564,819,207,303,340,847,894,502,572,032
2^{106} = 81,129,638,414,606,681,695,789,005,144,064
2^{107} = 162,259,276,829,213,363,391,578,010,288,128
2^{108} = 324,518,553,658,426,726,783,156,020,576,256
2^{109} = 649,037,107,316,853,453,566,312,041,152,512
2^{110} = 1,298,074,214,633,706,907,132,624,082,305,024
2^{111} = 2,596,148,429,267,413,814,265,248,164,610,048
2^{112} = 5,192,296,858,534,827,628,530,496,329,220,096 = 5.192297×10^{33}
2^{113} = 10,384,593,717,069,655,257,060,992,658,440,192
2^{114} = 20,769,187,434,139,310,514,121,985,316,880,384
2^{115} = 41,538,374,868,278,621,028,243,970,633,760,768
2^{116} = 83,076,749,736,557,242,056,487,941,267,521,536
2^{117} = 166,153,499,473,114,484,112,975,882,535,043,072
2^{118} = 332,306,998,946,228,968,225,951,765,070,086,144
2^{119} = 664,613,997,892,457,936,451,903,530,140,172,288
2^{120} = 1,329,227,995,784,915,872,903,807,060,280,344,576 = 1.329228×10^{36}
2^{121} = 2,658,455,991,569,831,745,807,614,120,560,689,152
2^{122} = 5,316,911,983,139,663,491,615,228,241,121,378,304
2^{123} = 10,633,823,966,279,326,983,230,456,482,242,756,608
2^{124} = 21,267,647,932,558,653,966,460,912,964,485,513,216 = 2.126765×10^{37}
2^{125} = 42,535,295,865,117,307,932,921,825,928,971,026,432
2^{126} = 85,070,591,730,234,615,865,843,651,857,942,052,864 = 8.507059×10^{37}
2^{127} = 170,141,183,460,469,231,731,687,303,715,884,105,728 = 1.701412×10^{38}
2^{128} = 340,282,366,920,938,463,463,374,607,431,768,211,456
2^{129} = 680,564,733,841,876,926,926,749,214,863,536,422,912
2^{130} = 1,361,129,467,683,753,853,853,498,429,727,072,845,824
2^{131} = 2,722,258,935,367,507,707,706,996,859,454,145,691,648
2^{132} = 5,444,517,870,735,015,415,413,993,718,908,291,383,296
2^{133} = 10,889,035,741,470,030,830,827,987,437,816,582,766,592
2^{134} = 21,778,071,482,940,061,661,655,974,875,633,165,533,184
2^{135} = 43,556,142,965,880,123,323,311,949,751,266,331,066,368
2^{136} = 87,112,285,931,760,246,646,623,899,502,532,662,132,736

$2^{137} =$ 174,224,571,863,520,493,293,247,799,005,065,324,265,472

$2^{138} =$ 348,449,143,727,040,986,586,495,598,010,130,648,530,944

$2^{139} =$ 696,898,287,454,081,973,172,991,196,020,261,297,061,888 $= 6.968983 \times 10^{41}$

$2^{140} =$ 1,393,796,574,908,163,946,345,982,392,040,522,594,123,776

$2^{141} =$ 2,787,593,149,816,327,892,691,964,784,081,045,188,247,552

$2^{142} =$ 5,575,186,299,632,655,785,383,929,568,162,090,376,495,104

$2^{143} =$ 11,150,372,599,265,311,570,767,859,136,324,180,752,990,208 $= 1.1 \times 10^{43}$

$2^{144} =$ 22,300,745,198,530,623,141,535,718,272,648,361,505,980,416

$2^{145} =$ 44,601,490,397,061,246,283,071,436,545,296,723,011,960,832

$2^{146} =$ 89,202,980,794,122,492,566,142,873,090,593,446,023,921,664

$2^{147} =$ 178,405,961,588,244,985,132,285,746,181,186,892,047,843,328 $= 1.784 \times 10^{44}$

$2^{148} =$ 356,811,923,176,489,970,264,571,492,362,373,784,095,686,656

$2^{149} =$ 713,623,846,352,979,940,529,142,984,724,747,568,191,373,312

$2^{150} =$ 1,427,247,692,705,959,881,058,285,969,449,495,136,382,746,624

$2^{151} =$ 2,854,495,385,411,919,762,116,571,938,898,990,272,765,493,248

$2^{152} =$ 5,708,990,770,823,839,524,233,143,877,797,980,545,530,986,496

$2^{153} =$ 11,417,981,541,647,679,048,466,287,755,595,961,091,061,972,992 $= 1.14 \times 10^{46}$

$2^{154} =$ 22,835,963,083,295,358,096,932,575,511,191,922,182,123,945,984

$2^{155} =$ 45,671,926,166,590,716,193,865,151,022,383,844,364,247,891,968

$2^{156} =$ 91,343,852,333,181,432,387,730,302,044,767,688,728,495,783,936

$2^{157} =$ 182,687,704,666,362,864,775,460,604,089,535,377,456,991,567,872

$2^{158} =$ 365,375,409,332,725,729,550,921,208,179,070,754,913,983,135,744

$2^{159} =$ 730,750,818,665,451,459,101,842,416,358,141,509,827,966,271,488

$2^{160} =$ 1,461,501,637,330,902,918,203,684,832,716,283,019,655,932,542,976
 $= 1.4615 \times 10^{48}$

$2^{161} =$ 2,923,003,274,661,805,836,407,369,665,432,566,039,311,865,085,952

$2^{162} =$ 5,846,006,549,323,611,672,814,739,330,865,132,078,623,730,171,904

$2^{163} =$ 11,692,013,098,647,223,345,629,478,661,730,264,157,247,460,343,808

$2^{164} =$ 23,384,026,197,294,446,691,258,957,323,460,528,314,494,920,687,616

$2^{165} =$ 46,768,052,394,588,893,382,517,914,646,921,056,628,989,841,375,232

$2^{166} =$ 93,536,104,789,177,786,765,035,829,293,842,113,257,979,682,750,464

$2^{167} =$ 187,072,209,578,355,573,530,071,658,587,684,226,515,959,365,500,928

$2^{168} =$ 374,144,419,156,711,147,060,143,317,175,368,453,031,918,731,001,856

$2^{169} =$ 748,288,838,313,422,294,120,286,634,350,736,906,063,837,462,003,712

$2^{170} =$ 1,496,577,676,626,844,588,240,573,268,701,473,812,127,674,924,007,424
 $= 1.496578 \times 10^{51}$

$2^{171} =$ 2,993,155,353,253,689,176,481,146,537,402,947,624,255,349,848,014,848

$2^{172} =$ 5,986,310,706,507,378,352,962,293,074,805,895,248,510,699, 696,029,696

$2^{173} =$ 11,972,621,413,014,756,705,924,586,149,611,790,497,021,399,392,059,392

$2^{174} =$ 23,945,242,826,029,513,411,849,172,299,223,580,994,042,798,784,118,784

$2^{175} =$ 47,890,485,652,059,026,823,698,344,598,447,161,988,085,597,568,237,568

$2^{176} =$ 95,780,971,304,118,053,647,396,689,196,894,323,976,171,195,136,475,136

$2^{177} =$ 191,561,942,608,236,107,294,793,378,393,788,647,952,342,390,272,950,272
 $= 1.9156194261 \times 10^{53}$

$2^{178} =$ 383,123,885,216,472,214,589,586,756,787,577,295,904,684,780,545,900,544

$2^{179} =$ 766,247,770,432,944,429,179,173,513,575,154,591,809,369,561,091,801,088

$2^{180} =$ 1,532,495,540,865,888,858,358,347,027,150,309,183,618,739,122,183,602,176
 $= 1.53249554 \times 10^{54}$

$2^{181} =$ 3,064,991,081,731,777,716,716,694,054,300,618,367,237,478,244,367,204,352

218

$2^{182} = $ 6,129,982,163,463,555,433,433,388,108,601,236,734,474,956,488,734,408,704

$2^{183} = $ 12,259,964,326,927,110,866,866,776,217,202,473,468,949,912,977,468,817,408

$2^{184} = $ 24,519,928,653,854,221,733,733,552,434,404,946,937,899,825,954,937,634,816

$2^{185} = $ 49,039,857,307,708,443,467,467,104,868,809,893,875,799,651,909,875,269,632

$2^{186} = $ 98,079,714,615,416,886,934,934,209,737,619,787,751,599,303,819,750,539,264

$2^{187} = $ 196,159,429,230,833,773,869,868,419,475,239,575,503,198,607,639,501,078,528

$2^{188} = $ 392,318,858,461,667,547,739,736,838,950,479,151,006,397,215,279,002,157, 056 = 3.923189 × 10^{56}

$2^{189} = $ 784,637,716,923,335,095,479,473,677,900,958,302,012,794,430,558,004,314, 112

$2^{190} = $ 1,569,275,433,846,670,190,958,947,355,801,916,604,025,588,861,116,008,628 ,224

$2^{191} = $ 3,138,550,867,693,340,381,917,894,711,603,833,208,051,177,722,232,017,256,448

$2^{192} = $ 6,277,101,735,386,680,763,835,789,423,207,666,416,102,355,444,464,034,512,896 = 6.27710736 × 10^{57}

$2^{193} = $ 12,554,203,470,773,361,527,671,578,846,415,332,832,204,710,888,928,069,025,792 = 1.25542 × 10^{58}

$2^{194} = $ 25,108,406,941,546,723,055,343,157,692,830,665,664,409,421,777,856,138,051,584

$2^{195} = $ 50,216,813,883,093,446,110,686,315,385,661,331,328,818,843,555,712,276,103,168

$2^{196} = $ 100,433,627,766,186,892,221,372,630,771,322,662,657,637,687,111,424,552,206,336 = 1.0043363 × 10^{59}

$2^{197} = $ 200,867,255,532,373,784,442,745,261,542,645,325,315,275,374,222,849,104,412,672

$2^{198} = $ 401,734,511,064,747,568,885,490,523,085,290,650,630,550,748,445,698,208,825,344

$2^{199} = $ 803,469,022,129,495,137,770,981,046,170,581,301,261,101,496,891,396,417,650,688

$2^{200} = $ 1,606,938,044,258,990,275,541,962,092,341,162,602,522,202,993,782,792,835,301,3 76 = 1.606938 × 10^{60}

$2^{201} = $ 3,213,876,088,517,980,551,083,924,184,682,325,205,044,405,987,565,585,670,602,7 52

$2^{202} = $ 6,427,752,177,035,961,102,167,848,369,364,650,410,088,811,975,131,171,341,205,5 04

$2^{203} = $ 12,855,504,354,071,922,204,335,696,738,729,300,820177,623,950,262,342,682,411,0 08 = 1.28555 × 10^{61}

$2^{204} = $ 25,711,008,708,143,844,408,671,393,477,458,601,640,355,247,900,524,685,364,822, 016

$2^{205} = $ 51,422,017,416,287,688,817,342,786,954,917,203,280,710,495,801,049,370,729,644, 032

$2^{206} = $ 102,844,034,832,575,377,634,685,573,909,834,406,561,420,991,602,098,741,459,288 ,064 = 1.02844 × 10^{62}

$2^{207} = $ 205,688,069,665,150,755,269,371,147,819,668,813,122,841,983,204,197,482,918,576 ,128

$2^{208} = $ 411,376,139,330,301,510,538,742,295,639,337,626,245,683,966,408,394,965,837,152 ,256

$2^{209} = $ 822,752,278,660,603,021,077,484,591,278,675,252,491,367,932,816,789,931,674,304 ,512

$2^{210} = $ 1,645,504,557,321,206,042,154,969,182,557,350,504,982,735,865,633,579,863,348,6 09,024 = 1.645505 × 10^{63}

$2^{211} = $ 3,291,009,114,642,412,084,309,938,365,114,701,009,965,471,731,267,159,726,697,2 18,048

$2^{212} = $ 6,582,018,229,284,824,168,619,876,730,229,402,019,930,943,462,534,319,453,394,4 36,096

$2^{213} = $ 13,164,036,458,569,648,337,239,753,460,458,804,039,861,886,925,068,638,906,788,

$872,192 = 1.316404 \times 10^{64}$

$2^{214} = 26,328,072,917,139,296,674,479,506,920,917,608,079,723,773,850,137,277,813,577,$
$744,384$

$2^{215} = 52,656,145,834,278,593,348,959,013,841,835,216,159,447,547,700,274,555,627,155,$
$488,768$

$2^{216} = 105,312,291,668,557,186,697,918,027,683,670,432,318,895,095,400,549,111,254,310$
$,977,536 = 1.053123 \times 10^{65}$

$2^{217} = 210,624,583,337,114,373,395,836,055,367,340,864,637,790,190,801,098,222,508,621$
$,955,072$

$2^{218} = 421,249,166,674,228,746,791,672,110,734,681,729,275,580,381,602,196,445,017,243$
$,910,144$

$2^{219} = 842,498,333,348,457,493,583,344,221,469,363,458,551,160,763,204,392,890,034,487$
$,820,288$

$2^{220} = 1,684,996,666,696,914,987,166,688,442,938,726,917,102,321,526,408,785,780,068,9$
$75,640,576 = 1.684997 \times 10^{66}$

$2^{221} = 3,369,993,333,393,829,974,333,376,885,877,453,834,204,643,052,817,571,560,137,9$
$51,281,152$

$2^{222} = 6,739,986,666,787,659,948,666,753,771,754,907,668,409,286,105,635,143,120,275,9$
$02,562,304$

$2^{223} = 13,479,973,333,575,319,897,333,507,543,509,815,336,818,572,211,270,286,240,551,$
$805,124,608 = 1.3479973 \times 10^{67}$

$2^{224} = 26,959,946,667,150,639,794,667,015,087,019,630,673,637,144,422,540,572,481,103,$
$610,249,216$

$2^{225} = 53,919,893,334,301,279,589,334,030,174,039,261,347,274,288,845,081,144,962,207,$
$220,498,432$

$2^{226} = 107,839,786,668,602,559,178,668,060,348,078,522,694,548,577,690,162,289,924,414$
$,440,996,864 = 1.0783979 \times 10^{68}$

$2^{227} = 215,679,573,337,205,118,357,336,120,696,157,045,389,097,155,380,324,579,848,828$
$,881,993,728$

$2^{228} = 431,359,146,674,410,236,714,672,241,392,314,090,778,194,310,760,649,159,697,657$
$,763,987,456$

$2^{229} = 862,718,293,348,820,473,429,344,482,784,628,181,556,388,621,521,298,319,395,315$
$,527,974,912$

$2^{230} = 1,725,436,586,697,640,946,858,688,965,569,256,363,112,777,243,042,596,638,790,6$
$31,055,949,824 = 1.7254359 \times 10^{69}$

$2^{231} = 3,450,873,173,395,281,893,717,377,931,138,512,726,225,554,486,085,193,277,581,2$
$62,111,899,648$

$2^{232} = 6,901,746,346,790,563,787,434,755,862,277,025,452,451,108,972,170,386,555,162,5$
$24,223,799,296$

$2^{233} = 13,803,492,693,581,127,574,869,511,724,554,050,904,902,217,944,340,773,110,325,$
$048,447,598,592 = 1.3803493 \times 10^{70}$

$2^{234} = 27,606,985,387,162,255,149,739,023,449,108,101,809,804,435,888,681,546,220,650,$
$096,895,197,184$

$2^{235} = 55,213,970,774,324,510,299,478,046,898,216,203,619,608,871,777,363,092,441,300,$
$193,790,394,368$

$2^{236} = 110,427,941,548,649,020,598,956,093,796,432,407,239,217,743,554,726,184,882,600$
$,387,580,788,736 = 1.1042794 \times 10^{71}$

$2^{237} = 220,855,883,097,298,041,197,912,187,592,864,814,478,435,487,109,452,369,765,200$
$,775,161,577,472$

$2^{238} =$ 441,711,766,194,596,082,395,824,375,185,729,628,956,870,974,218,904,739,530,401 ,550,323,154,944

$2^{239} =$ 883,423,532,389,192,164,791,648,750,371,459,257,913,741,948,437,809,479,060,803 ,100,646,309,888

$2^{240} =$ 1,766,847,064,778,384,329,583,297,500,742,918,515,827,483,896,875,618,958,121,6 06,201,292,619,776 = 1.7668471×10^{72}

$2^{241} =$ 3,533,694,129,556,768,659,166,595,001,485,837,031,654,967,793,751,237,916,243,2 12,402,585,239,552

$2^{242} =$ 7,067,388,259,113,537,318,333,190,002,971,674,063,309,935,587,502,475,832,486,4 24,805,170,479,104

$2^{243} =$ 14,134,776,518,227,074,636,666,380,005,943,348,126,619,871,175,004,951,664,972, 849,610,340,958,208 = 1.4134777×10^{73}

$2^{244} =$ 28,269,553,036,454,149,273,332,760,011,886,696,253,239,742,350,009,903,329,945, 699,220,681,916,416

$2^{245} =$ 56,539,106,072,908,298,546,665,520,023,773,392,506,479,484,700,019,806,659,891, 398,441,363,832,832

$2^{246} =$ 113,078,212,145,816,597,093,331,040,047,546,785,012,958,969,400,039,613,319,782 ,796,882,727,665,664 = 1.1307821×10^{74}

$2^{247} =$ 226,156,424,291,633,194,186,662,080,095,093,570,025,917,938,800,079,226,639,565 ,593,765,455,331,328

$2^{248} =$ 452,312,848,583,266,388,373,324,160,190,187,140,051,835,877,600,158,453,279,131 ,187,530,910,662,656

$2^{249} =$ 904,625,697,166,532,776,746,648,320,380,374,280,103,671,755,200,316,906,558,262 ,375,061,821,325,312

$2^{250} =$ 1,809,251,394,333,065,553,493,296,640,760,748,560,207,343,510,400,633,813,116,5 24,750,123,642,650,624 = 1.8092514×10^{75}

$2^{251} =$ 3,618,502,788,666,131,106,986,593,281,521,497,120,414,687,020,801,267,626,233,0 49,500,247,285,301,248

$2^{252} =$ 7,237,005,577,332,262,213,973,186,563,042,994,240,829,374,041,602,535,252,466,0 99,000,494,570,602,496

$2^{253} =$ 14,474,011,154,664,524,427,946,373,126,085,988,481,658,748,083,205,070,504,932, 198,000,989,141,204,992 = 1.4474011×10^{76}

$2^{254} =$ 28,948,022,309,329,048,855,892,746,252,171,976,963,317,496,166,410,141,009,864, 396,001,978,282,409,984

$2^{255} =$ 57,896,044,618,658,097,711,785,492,504,343,953,926,634,992,332,820,282,019,728, 792,003,956,564,819,968

$2^{256} =$ 115,792,089,237,316,195,423,570,985,008,687,907,853,269,984,665,640,564,039,457 ,584,007,913,129,639,936 = 1.1579209×10^{77}

$2^{257} =$ 231,584,178,474,632,390,847,141,970,017,375,815,706,539,969,331,281,128,078,915 ,168,015,826,259,279,872 = 2.3158418×10^{77}.

$2^{264} =$ 29,642,774,844,752,946,028,434,172,162,224,104,410,437,116,074,403,984,394,101, 141,506,025,761,187,823,616 = 2.9642775×10^{79}

$2^{265} =$ 59,285,549,689,505,892,056,868,344,324,448,208,820,874,232,148,807,968,788,202, 283,012,051,522,375,647,232

$2^{266} =$ 118,571,099,379,011,784,113,736,688,648,896,417,641,748,464,297,615,937,576,404 ,566,024,103,044,751,294,464 = 1.1845711×10^{80}

$2^{274} =$ 30,354,201,441,027,016,733,116,592,294,117,482,916,287,606,860,189,680,019,559, 568,902,170,379,456,331,382,784 = 3.0354201×10^{82}

$2^{275} =$ 60,708,402,882,054,033,466,233,184,588,234,965,832,575,213,720,379,360,039,119,

221

137,804,340,758,912,662,765,568

$2^{276} =$ 121,416,805,764,108,066,932,466,369,176,469,931,665,150,427,440,758,720,078,238,275,608,681,517,825,325,531,136 = $1.21416806 \times 10^{83}$

$2^{284} =$ 31,082,702,275,611,665,134,711,390,509,176,302,506,278,509,424,834,232,340,028,998,555,822,468,563,283,335,970,816 = 3.1082702×10^{85}

$2^{285} =$ 62,165,404,551,223,330,269,422,781,018,352,605,012,557,018,849,668,464,680,057,997,111,644,937,126,566,671,941,632

$2^{286} =$ 124,330,809,102,446,660,538,845,562,036,705,210,025,114,037,699,336,929,360,115,994,223,289,874,253,133,343,883,264 = 1.2833081×10^{86}

$2^{294} =$ 31,828,687,130,226,345,097,944,463,881,396,533,766,429,193,651,030,253,916,189,694,521,162,207,808,802,136,034,115,584 = 3.1828687×10^{88}

$2^{295} =$ 63,657,374,260,452,690,195,888,927,762,793,067,532,858,387,302,060,507,832,379,389,042,324,415,617,604,272,068,231,168

$2^{296} =$ 127,314,748,520,905,380,391,777,855,525,586,135,065,716,774,604,121,015,664,758,778,084,648,831,235,208,544,136,462,336 = 1.2731475×10^{89}

$2^{304} =$ 32,592,575,621,351,777,380,295,131,014,550,050,576,823,494,298,654,980,010,178,247,189,670,100,796,213,387,298,934,358,016 = 3.8321796×10^{91}

$2^{305} =$ 65,185,151,242,703,554,760,590,262,029,100,101,153,646,988,597,309,960,020,356,494,379,340,201,592,426,774,597,868,716,032

$2^{306} =$ 130,370,302,485,407,109,521,180,524,058,200,202,307,293,977,194,619,920,040,712,988,758,680,403,184,853,549,195,737,432,064 = 1.3037030×10^{92}

$2^{314} =$ 33,374,797,436,264,220,037,422,214,158,899,251,790,667,258,161,822,699,530,422,525,122,222,183,215,322,508,594,108,782,608,384 = 3.3374797×10^{94}

$2^{315} =$ 66,749,594,872,528,440,074,844,428,317,798,503,581,334,516,323,645,399,060,845,050,244,444,366,430,645,017,188,217,565,216,768

$2^{316} =$ 133,499,189,745,056,880,149,688,856,635,597,007,162,669,032,647,290,798,121,690,100,488,888,732,861,290,034,376,435,130,433,536 = 1.3349919×10^{95}

$2^{324} =$ 34,175,792,574,734,561,318,320,347,298,712,833,833,643,272,357,706,444,319,152,665,725,155,515,612,490,248,800,367,393,390,985,216 = 3.4175793×10^{97}

$2^{325} =$ 68,351,585,149,469,122,636,640,694,597,425,667,667,286,544,715,412,888,638,305,331,450,311,031,224,980,497,600,734,786,781,970,432

$2^{326} =$ 136,703,170,298,938,245,273,281,389,194,851,335,334,573,089,430,825,777,276,610,662,900,622,062,449,960,995,201,469,573,563,940,864 = 1.3670317×10^{98}

$2^{334} =$ 34,996,011,596,528,190,789,960,035,633,881,941,845,650,710,894,291,398,982,812,329,702,559,247,987,190,014,771,576,210,832,368,861,184 = $3.4996012 \times 10^{100}$

$2^{335} =$ 69,992,023,193,056,381,579,920,071,267,763,883,691,301,421,788,582,797,965,624,659,405,118,495,974,380,029,543,152,421,664,737,722,368

$2^{336} =$ 139,984,046,386,112,763,159,840,142,535,527,767,382,602,843,577,165,595,931,249,318,810,236,991,948,760,059,086,304,843,329,475,444,736 = $1.3998405 \times 10^{101}$

$2^{344} =$ 35,835,915,874,844,867,368,919,076,489,095,108,449,946,327,955,754,392,558,399,825,615,420,669,938,882,575,126,094,039,892,345,713,852,416 = $3.5835916 \times 10^{103}$

$2^{345} =$ 71,671,831,749,689,734,737,838,152,978,190,216,899,892,655,911,508,785,116,799,651,230,841,339,877,765,150,252,188,079,784,691,427,704,832

$2^{346} =$ 143,343,663,499,379,469,475,676,305,956,380,433,799,785,311,823,017,570,233,599,302,461,682,679,755,530,300,504,376,159,569,382,855,409,664 = $1.4334366 \times 10^{104}$.

$2^{354} =$ 36,695,977,855,841,144,185,773,134,324,833,391,052,745,039,826,692,497,979,801,

421,430,190,766,017,415,756,929,120,296,849,762,010,984,873,984 = 3.6695978 × 10^{106}

$2^{355} =$ 73,391,955,711,682,288,371,546,268,649,666,782,105,490,079,653,384,995,959,602,842,860,381,532,034,831,513,858,240,593,699,524,021,969,747,968

$2^{356} =$ 146,783,911,423,364,576,743,092,537,299,333,564,210,980,159,306,769,991,919,205,685,720,763,064,069,663,027,716,481,187,399,048,043,939,495,936 = 1.4678391 × 10^{107}

$2^{364} =$ 37,576,681,324,381,331,646,231,689,548,629,392,438,010,920,782,533,117,931,316,655,544,515,344,401,833,735,095,419,183,974,156,299,248,510,959,616 = 3.7576813 × 10^{109}

$2^{365} =$ 75,153,362,648,762,663,292,463,379,097,258,784,876,021,841,565,066,235,862,633,311,089,030,688,803,667,470,190,838,367,948,312,598,497,021,919,232

$2^{366} =$ 150,306,725,297,525,326,584,926,758,194,517,569,752,043,683,130,132,471,725,266,622,178,061,377,607,334,940,381,676,735,896,625,196,994,043,838,464 = 1.5030673 × 10^{110}

$2^{374} =$ 38,478,521,676,166,483,605,741,250,097,796,497,856,523,182,881,313,912,761,668,255,277,583,712,667,477,744,737,709,244,389,536,050,430,475,222,646,784 = 3.8478522 × 10^{112}

$2^{375} =$ 76,957,043,352,332,967,211,482,500,195,592,995,713,046,365,762,627,825,523,336,510,555,167,425,334,955,489,475,418,488,779,072,100,860,950,445,293,568

$2^{376} =$ 153,914,086,704,665,934,422,965,000,391,185,991,426,092,731,525,255,651,046,673,021,110,334,850,669,910,978,950,836,977,558,144,201,721,900,890,587,136 = 1.5391409 × 10^{113}

$2^{384} =$ 39,402,006,196,394,479,212,279,040,100,143,613,805,079,739,270,465,446,667,948,293,404,245,721,771,497,210,611,414,266,254,884,915,640,806,627,990,306,816 = 3.9402006 × 10^{115}.

$2^{385} =$ 78,804,012,392,788,958,424,558,080,200,287,227,610,159,478,540,930,893,335,896,586,808,491,443,542,994,421,222,828,532,509,769,831,281,613,255,980,613,632 = 7.8804012 × 10^{115}

$2^{512} =$ 1.340781 × 10^{154} = (2^{256})^2
$2^{769} =$ 3.1050 × 10^{231} = 2[(2^{256})^3]

✝✡✝✡✝✡✝✡✝✡✝✡✝✡✝✡✝✡✝✡✝✡✝✡✝✡

But some have maintained that Metaphysics, considered as a quest for real explanations, is a fool's errand. Kant, writing in 18th Century Germany, labelled it a "transcendental illusion." Since Kant, many thinkers have taken it for granted that Metaphysics can at best achieve a subjectively convenient construction—a schematization—of the pattern of reality. To ask it to provide an objective grasp of the way things really are is to demand the impossible.
——**AVERY DULLES, S. J.; JAMES M. DEMSKE, S. J.; & ROBERT J. O'CONNELL, S. J.:** *Introductory Metaphysics* as given on page 9 published by Sheed & Ward, New York, 1955. [If Esoptrics is correct, the "many thinkers" described above shall find themselves with a great deal of egg on their faces. For, if Esoptrics is correct, it clearly proves that *only* Metaphysics can "provide an objective grasp of the way things really are".]

APPENDIX C:

POWERS OF 6

2	36
3	216
4	1,296
5	7,776
6	46,656
7	279,936
8	1,679,616
9	10,077,696
10	60,466,176
11	362,797,056
12	2,176,782,336
13	13,060,694,016
14	78,364,164,096
15	470,184,984,576
16	2,821,109,907,456 = 2.82111 x 10^{12}.
17	16,926,659,444,736
18	101,559,956,668,416
19	609,359,740,010.496
20	3,656,158,440,062,976
21	21,936,950,640,377,856
22	131,621,703,842,267,136
23	789,730,223,053,602,816
24	4,738,381,338,321,616,896
25	28,430,288,029,929,701,376
26	170,581,728,179,578,208,256
27	1,023,490,369,077,469,249,536
28	6,140,942,214,464,815,497,216
29	36,845,653,286,788,892,983,296
30	221,073,919,720,733,357,899,776
31	1,326,443,518,324,400,147,398,656
32	7,958,661,109,946,400,884,391,936 = 7.95866111 x 10^{24}
33	47,751,966,659,678,405,306,351,616
34	286,511,799,958,070,431,838,109,696
35	1,719,070,799,748,422,591,028,658,176
36	10,314,424,798,490,535,546,171,949,056
37	61,886,548,790,943,213,277 031,694,336
38	371,319,292,745,659,279,662,190,166,016

39	2,227,915,756,473,955,677,973,140,996,096
40	13,367,494,538,843,734,067,838,845,976,576
41	80,204,967,233,062,404,407,033,075,859,456
42	481,229,803,398,374,426,442,198,455,156,736
43	2,887,378,820,390,246,558,653,190,730,940,416
44	17,324,272,922,341,479,351,919,144,385,642,496
45	103,945,637,534,048,876,111,514,866,313,854,976
46	623,673,825,204,293,256,669,089,197,883,129,856
47	3,742,042,951,225,759,540,014,535,187,298,779,136
48	22,452,257,707,354,557,240,087,211,123,792,674,816
49	134,713,546,244,127,343,440,523,266,742,756,048,896
50	808, 281,277,464,764,060,643,139,600,456,536,293,376
51	4,849,687,664,788,584,363,858,837,602,739,217,760,256
52	29,098,125,988,731,506,183,153,025,616,435,306,561,536
53	174,588,755,932,389,037,098,918,153,698,611,839,369,216
54	1,047,532,535,594,334,222,593,508,922,191,671,036,215,296
55	6,285,195,213,566,005,335,561,053,533,150,026,217,291,776
56	37,711,171,281,396,032,013,366,321,198,900,157,303,750,656
57	226,267,027,688,376,192,080,197,927,193,400,943,822,503,936
58	1,357,602,166,130,257,152,481,187,563,160,405,662,935,023,616
59	8,145,612,996,781,542,914,887,125,378,962,433,977,610,141,696
60	48,873,677,980,689,257,489,322,752,273,774,603,865,660,850,176
61	293,242,067,884,135,544,935,936,513,642,647,623,193,965,101,056
62	1,759,452,407,304,813,269,615,619,081,855,885,739,163,790,606,336
63	10,556,714,443,828,879,617,693,714,491,135,314,434,982,743,638,016
64	63,340,286,662,973,277,706,162,286,946,811,886,609,896,461,828,096
64	$63.34028666 \times 10^{48} = 6.334028666 \times 10^{49} = (7.95866111 \times 10^{24})^2$
128	$4,011.992 \times 10^{96} = 4.011992 \times 10^{99} = (63.34029 \times 10^{48})^2$
256	$16.09608 \times 10^{198} = 1.61 \times 10^{199} = (4.011992 \times 10^{99})^2$

APPENDIX D:

LB&D OF
SELECT ONTOLOGICAL DISTANCES
(For the LB&D in Φ, double the ontological distance.)

LB&D OF ACTUALITY DURING FIRST HALF OF FORMS CYCLE

AT	CENTIMETERS	MILES
2^{128}	0.00 00 00 05 $= 2^{129}\Phi$	
2^{129}	0.00 00 00 10^1 $= 2^{130}\Phi$	
2^{130}	0.00 00 00 20 $= 2^{131}\Phi$	
2^{131}	0.00 00 00 40 $= 2^{132}\Phi$	
2^{132}	0.00 00 00 80 $= 2^{133}\Phi$	
2^{133}	0.00 00 01 60	
2^{134}	0.00 00 03 20	
2^{135}	0.00 00 06 40	
2^{136}	0.00 00 12 80	
2^{137}	0.00 00 25 60	
2^{138}	0.00 00 51 20	
2^{139}	0.00 01 02 40	
2^{140}	0.00 02 04 80	
2^{141}	0.00 04 09 60	
2^{142}	0.00 08 19 20	
2^{143}	0.00 16 38 40	
2^{144}	0.00 32 76 80	
2^{145}	0.00 65 53 60	
2^{146}	0.01 31 07 20	
2^{147}	0.02 62 14 40	
2^{148}	0.05 24 28 80	
2^{149}	0.10 48 57 60	
2^{150}	0.20 97 15 20	
2^{151}	0.41 94 30 40	
2^{152}	0.83 88 60 80	
2^{153}	1.67 77 21 60	
2^{154}	3.35 54 43 20	
2^{155}	6.71 08 86 40 $= 2^{156}\Phi$	
2^{156}	13.42 17 72 80 $= 2^{157}\Phi$	
2^{157}	26.84 35 45 60	
2^{158}	53.68 70 91 20	
2^{159}	107.37 41 82 40	

---

^{1}From this line onward, the figure in this column is merely the previous line multiplied by 2.

AT	LB&D OF ACTUALITY DURING FIRST HALF OF FORM'S CYCLE	
	CENTIMETERS	MILES
2^{160}	214.74 83 64 80 ÷ 160,934.72 =	0.00 13 34 38 18 213
2^{161}	429.49 67 29 60 ÷ 160,934.72 =	0.00 26 68 76 36 43
2^{162}	858.99 34 59 20 ÷ 160,934.72 =	0.00 53 37 52 72 85
2^{163}	1,717.98 69 18 40 ÷ 160,934.72 =	0.01 06 75 05 45 71
2^{164}	3,435.97 38 36 80 ÷ 160,934.72 =	0.02 13 50 10 91 42
2^{165}	6,871.94 76 73 60 ÷ 160,934.72 =	0.04 27 00 21 82 84
2^{166}	13,743.89 53 47 20 ÷ 160,934.72 =	0.08 54 00 43 65 70
2^{167}	27,487.79 06 94 40 ÷ 160,934.72 =	0.17 08 00 87 31 39
2^{168}	54,975.58 13 88 80 ÷ 160,934.72 =	0.34 16 01 74 62 78
2^{169}	109,951.16 27 77 60 ÷ 160,934.72 =	0.68 32 03 49 25 56
2^{170}	219,902.32 55 55 20 ÷ 160,934.72 =	1.36 64 06 98 51 13
2^{171}	439,804.65 11 10 40 ÷ 160,934.72 =	2.73 28 13 97 02 26
2^{172}	879,609.30 22 20 80 ÷ 160,934.72 =	5.46 56 27 94 04 52
2^{173}	1,759,218.60 44 41 60 ÷ 160,934.72 =	10.93 12 55 88 09 03
2^{174}	3,518,437.20 88 83 20 ÷ 160,934.72 =	21.86 25 11 76 18 07
2^{175}	7,036,874.41 77 66 40 ÷ 160,934.72 =	43.72 50 23 52 36 15
2^{176}	14,073,748.83 55 32 80	87.45 00 47 04 72 30[1]
2^{177}	28,147,497.67 10 65 60	174.90 00 94 09 44 60
2^{178}	56,294,995.34 21 31 20	349.80 01 88 18 89 20
2^{179}	112,589,990.68 42 62 40= $2^{180}\Phi$	699.60 03 76 37 78 40
2^{180}	225,179,981.36 85 24 80	1,399.20 07 52 75 56 80
2^{181}	450,359,962.73 70 49 60	2,798.40 15 05 51 13 60
2^{182}	900,719,925.47 40 99 20	5,596.80 30 11 02 27 20
2^{183}	1,801,439,850.94 81 98 40	11,193.60 60 22 04 54 40
2^{184}	3,602,879,701.89 63 96 80	22,387.21 20 44 09 08 80
2^{185}	7,205,759,403.79 27 93 60	44,774.42 40 88 18 17 60
2^{186}	14,411,518,807.58 55 87 20	89,548.84 81 76 36 35 20
2^{187}	28,823,037,615.17 11 74 40 = 2.88×10^{10} cm.	179,097.69 63 52 72 70 40
2^{188}	57,646,075,230.34 23 48 80 = $2^{189}\Phi$	358,195.39 27 05 45 40 80
2^{189}	115,292,150,460.68 46 97 60	716,390.78 54 10 90 81 60
2^{190}	230,584,300,921.36 93 95 20	1,432,781.57 08 21 81 63 20
2^{191}	461,168,601,842.73 87 90 40	2,865,563.14 16 43 63 26 40
2^{192}	922,337,203,685.47 75 80 80	5,731,126.28 32 87 26 52 80
2^{193}	1,844,674,407,370.95 51 61 60	11,462,252.56 65 74 53 05 60
2^{194}	3,689,348,814,741.91 03 23 20	22,924,505.13 31 49 06 11 20
2^{195}	7,378,697,629,483.82 06 46 40	45,849,010.26 62 98 12 22 40
2^{196}	14,757,395,258,967.64 12 92 80	91,698,020.53 25 96 24 44 80
2^{197}	29,514,790,517,935.28 25 85 60	183,396,041.06 51 92 48 89 60
2^{198}	59,029,581,035,870.56 51 71 20	366,792,082.13 03 84 97 79 20
2^{199}	118,059,162,071,741.13 03 42 40	733,584,164.26 07 69 95 58 40
2^{200}	236,118,324,143,482.26 06 84 80	1,467,168,328.52 15 39 91 16 80
2^{201}	472,236,648,286,964.52 13 69 60	2,934,336,657.04 30 79 82 33 60
2^{202}	944,473,296,573,929.04 27 39 20	5,868,673,314.08 61 59 64 67 20

[1]From this line onward, the figure in this column is merely the previous line multiplied by 2.

LB&D OF ACTUALITY DURING FIRST HALF OF FORM'S CYCLE

AT	CENTIMETERS	MILES
2^{203}	1,888,946,593,147,858.08 54 78 40	11,737,346,628.17 23 19 29 34 40
2^{204}	3,777,893,186,295,716.17 09 56 80	23,474,693,256.34 46 38 58 68 80
2^{205}	7,555,786,372,591,432.34 19 13 60	46,949,386,512.68 92 77 17 37 60
2^{206}	15,111,572,745,182,864.68 38 27 20	93,898,773,025.37 85 54 34 75 20
2^{207}	30,223,145,490,365,729.36 76 54 40	187,797,546,050.75 71 08 69 50 40
2^{208}	60,446,290,980,731,458.73 53 08 80	375,595,092,101.51 42 17 39 00 80
2^{209}	120,892,581,961,462,917.47 06 17 60	751,190,184,203.02 84 34 78 01 60
2^{210}	241,785,163,922,925,834.94 12 35 20	1,502,380,368,406.05 68 69 56 03 20
2^{211}	483,570,327,845,851,669.88 24 70 40	3,004,760,736,812.11 37 39 12 06 40
2^{212}	967,140,655,691,703,339.76 49 40 80	6,009,521,473,624.22 74 78 24 12 80 = 1.0222885 ly.[1]
2^{213}	1,934,281,311,383,406,679.52 98 81 60	12,019,042,947,248.45 49 56 48 25 60 = 2.044577 ly.
2^{214}	3,868,562,622,766,813,359.05 97 63 20	24,038,085,894,496.90 99 12 96 51 20 = 4.089154 ly.
2^{215}	7,737,125,245,533,626,718.11 95 26 40	48,076,171,788,993.81 98 25 93 02 40 = 8.178308 ly.
2^{216}	15,474,250,491,067,253,436.23 90 52 80	96,152,343,577,987.63 96 51 86 04 80 = 16.356616 ly.
2^{217}	30,948,500,982,134,506,872.47 81 05 60	192,304,687,155,975.27 93 03 72 09 60 = 32.713232 ly.
2^{218}	61,897,001,964,269,013,744.95 62 11 20	384,609,374,311,950.55 86 07 44 19 20 = 65.426 464 ly.
2^{219}	123,794,003,928,538,027,489.91 24 22 40	769,218,748,623,901.11 72 14 88 38 40 = 130.852928 ly.
2^{220}	247,588,007,857,076,054,979.82 48 44 80	1,538,437,497,247,802.23 44 29 76 76 80 = 261.705856 ly.
2^{221}	495,176,015,714,152,109,959.64 96 89 60	3,076,874,994,495,604.46 88 59 53 53 60 = 523.411712 ly.
2^{222}	990,352,031,428,304,219,919.29 93 79 20	6,153,749,988,991,208.93 77 19 07 07 20= 1,046.823424 ly.
2^{223}	1,980,704,062,856,608,439,838.59 87 58 40	12,307,499,977,982,417.87 54 38 14 14 40 = 2,093.646848 ly.
2^{224}	3,961,408,125,713,216,879,677.19 75 16 80	24,614,999,955,964,835.75 08 76 28 28 80 = 4,187.293696 ly.
2^{225}	7,922,816,251,426,433,759,354.39 50 33 60	49,229,999,911,929,671.50 17 52 56 57 60 = 8,374.587392 ly.
2^{226}	15,845,632,502,852,867,518,708.79 00 67 20	98,459,999,823,859,343.00 35 05 13 15 20= 16,749.174784 ly.
2^{227}	31,691,265,005,705,735,037,417.58 01 34	196,919,999,647,718,686.00 70 10 26 30

---

[1]Ly = the distance light travels in a tropical year of 31,556,925.9747 seconds, if the speed of light per second is 29,979,300,000 cm., which is to say 946,054,550,873,323,710 cm. per year = 9,460,545, 508,733.23710 km. per yr.. If 9,460,545,508,733.2 be divided by 1.6093472, then light travels 5,878,498,753,241.8 miles in a tropical year. If 6,009,521,473,624.2 miles be divided by 5,878,498, 753,241.8, then the LB&D of the actuality at 2^{212} is 1.022288466 light years.

LB&D OF ACTUALITY DURING FIRST HALF OF FORM'S CYCLE

AT	CENTIMETERS	MILES
	40	40 = 33,498.349568 ly.
2^{228}	63,382,530,011,411,470,074,835.16 02 68 80	393,839,999,295,437,372.01 40 20 52 60 80 = 66,996.699136 ly.
2^{229}	126,765,060,022,822,940,149,670.32 05 37 60	787,679,998,590,874,744.02 80 41 05 21 60 = 133,993.398272 ly.
2^{230}	253,530,120,045,645,880,299,340.64 10 75 20	1,575,359,997,181,749,488.05 60 82 10 43 20= 267,986.796544 ly.
2^{231}	507,060,240,091,291,760,598,681.28 21 50 40	3,150,719,994,363,498,976.11 21 64 20 86 40 = 535,973.593088 ly.
2^{232}	1,014,120,480,182,583,521,197,362.56 43 00 80	6,301,439,988,726,997,952.22 43 28 41 72 80= 1,071,947.186176 ly.
2^{233}	2,028,240,960,365,167,042,394,725.12 86 01 60	12,602,879,977,453,995,904.44 86 56 83 45 60 = 2,143,894.372352 ly.
2^{234}	4,056,481,920,730,334,084,789,450.25 72 03 20	25,205,759,954,907,991,808.89 73 13 66 91 20 = 4,287,788.744704 ly.
2^{235}	8,112,963,841,460,668,169,578,900.51 44 06 40	50,411,519,909,815,983,617.79 46 27 33 82 40 = 8,575,577.489408 ly.
2^{236}	16,225,927,682,921,336,339,157,801.02 88 12 80	100,823,039,819,631,967,235.58 92 54 67 64 80 = 17,151,154.978816 ly.
2^{237}	32,451,855,365,842,672,678,315,602.05 76 25 60	201,646,079,639,263,934,471.17 85 09 35 29 60 = 34,302,309.957632 ly.
2^{238}	64,903,710,731,685,345,356,631,204.11 52 51 20	403,292,159,278,527,868,942.35 70 18 70 59 20 = 68,604,619.915264 ly.
2^{239}	129,807,421,463,370,690,713,262,408.23 05 02 40	806,584,318,557,055,737,884.71 40 37 41 18 40 = 137,209,239.830528 ly.
2^{240}	259,614,842,926,741,381,426,524,816.46 10 04 80	1,613,168,637,114,111,475,769.42 80 74 82 36 80 = 274,418,479.661056 ly.
2^{241}	519,229,685,853,482,762,853,049,632.92 20 09 60	3,226,337,274,228,222,951,538.85 61 49 64 73 60 = 548,836,959.322112 ly.
2^{242}	1,038,459,371,706,965,525,706,099,265.8 4 40 19 20	6,452,674,548,456,445,903,077.71 22 99 29 47 20 = 1,097,673,918.644224 ly.
2^{243}	2,076,918,743,413,931,051,412,198,531.6 8 80 38 40	12,905,349,096,912,891,806,155.42 45 98 58 94 40 = 2,195,347,837.288448 ly.
2^{244}	4,153,837,486,827,862,102,824,397,063.3 7 60 76 80	25,810,698,193,825,783,612,310.84 91 97 17 88 80 = 4,390,695,674.576896 ly.
2^{245}	8,307,674,973,655,724,205,648,794,126.7 5 21 53 60	51,621,396,387,651,567,224,621.69 83 94 35 77 60 = 8,781,391,349.153792 ly.
2^{246}	16,615,349,947,311,448,411,297,588,253. 50 43 07 20	103,242,792,775,303,134,449,243.39 67 88 71 55 20 = 17,562,782,698.307584 ly.
2^{247}	33,230,699,894,622,896,822,595,176,507. 00 86 14 40	206,485,585,550,606,268,898,486.79 35 77 43 10 40

LB&D OF ACTUALITY DURING FIRST HALF OF FORM'S CYCLE

AT	CENTIMETERS	MILES
2^{248}	66,461,399,789,245,793,645,190,353,014.01 72 28 80	= 35,125,565,396.615168 ly. 412,971,171,101,212,537,796,973.58 71 54 86 20 80
2^{249}	132,922,799,578,491,587,290,380,706,028.03 44 57 60	= 70,251,130,793.23 03 36 ly. 825,942,342,202,425,075,593,947.17 43 09 72 41 60
2^{250}	265,845,599,156,983,174,580,761,412,056.06 89 15 20	= 140,502,261,586.460672 ly. 1,651,884,684,404,850,151,187,894.34 86 19 44 83 20
2^{251}	531,691,198,313,966,349,161,522,824,112.13 78 30 40	= 281,004,523,172.921344 ly. 3,303,769,368,809,700,302,375,788.69 72 38 89 66 40
2^{252}	1,063,382,396,627,932,698,323,045,648,224.27 56 60 80	= 562,009,046,345.842688 ly. 6,607,538,737,619,400,604,751,577.39 44 77 79 32 80
2^{253}	2,126,764,793,255,865,396,646,091,296,448.55 13 21 60	= 1,124,018,092,691.685376 ly. 13,215,077,475,238,801,209,503,154.78 89 55 58 65 60
2^{254}	4,253,529,586,511,730,793,292,182,592,897.10 26 43 20	= 2,248,036,185,383.370752 ly. 26,430,154,950,477,602,419,006,309.57 79 11 17 31 20
2^{255}	8,507,059,173,023,461,586,584,365,185,794.20 52 86 40	= 4,496,072,370,766.741504 ly. 52,860,309,900,955,204,838,012,619.15 58 22 34 62 40
2^{256}	17,014,118,346,046,923,173,168,730,371,588.41 05 72 80 = 1.7014118×10^{31}	= 8,992,144,741,533.483008 ly. 105,720,619,801,910,409,676,025,238.31 16 44 69 24 80 = 1.0572063×10^{26} = 17,984,289,483,066.966016 ly.

✝✡✝✡✝✡✝✡✝✡✝✡✝✡✝✡✝✡✝✡✝✡✝✡✝✡✝✡

Ad Majorem Christi Suaeque Ecclesiae Gloriam,
Consummatum Est
In Die Domini,
10/18/09

GENERAL INDEX

A.

ABSOLUTES: not a function of anything else: 192.

ABSTRACTABLE, PRIME: see: PRIME ABSTRACTABLE.

ABSTRACTIONS: not divisible in space in fact: 202-203.

ACCELERATION: above mirror threshold is mirrored below it: 48 ftn. #1, 182; caused by generators colliding logically: 37; causes lead form to ingest others: 37-38; means form moving to a current OD higher than its native one: 10, 87; limits of: 11, 179; slows time for forms but opposite for generators: 37, 87.

ACCELERATION, THE GREAT: brief mention: 11, 34; in detail: 38-43; significance of: 42-43.

ACHILLES: & the tortoise: 25.

ACTS, PRIMARY: also see POWERS, SIX; xv; absolute duration of cannot be measured by time: 24-25, 31-32, 77, 88-89, 100, 193-194; an act in no sense known to us: 3, 195; characterized by a level of power: 3, 195; equal acts of existence: 81, 127-128; fastest rate of change from one act to another is at OD1: 88; hereafter "act" always means "primary act": 3; in what sense a cube: 4, 21-23; for forms vs. those of generators: 4, 19-20; like God not involved in time while still the same one act: 24; Little Black Sambo's tigers as an example of: 195; no trace of change while the same one act: 81-82, 195; not observable by us: 81, 203; number of, available to a generator = $(2N_c)^3$: 23; most powerful is 2^{256} times the least powerful: 15, 16; their 6 modes of power: 4; something an ultimate *is* rather than merely ***does***: 130, 195; two per cycle in forms: 89; vs. secondary acts: 81-82, 195; we have not the slightest hint of what it's like to experience their internal characteristics: 203.

ACTS, SECONDARY: behavior produced by changes from primary act to primary act: 81, 128; only kind of acts we can observe: 81.

ALEPHONS: also see: MONONS; also see: DIONS; also see: SEXTONS; Aleph, Beth & Daleth variations: 153; each is one piggyback form & its carrying generator: 152; further divided into monons, dions, and sextons: 155-156; number of pos. vs. neg. kinds in universe: 152, 152 ftn. #1; same as Science's neutrinos & antineutrinos: 152 ftn. #1; three kinds of pos. & 3 kinds of neg.: 152-153.

ALEX: scientist at a major university: 185; sent me a copy of a lecture by Sir Michael Atiyah & tells me Atiyah refers to the mismatch between special relativity and quantum physics: 185.

ALGEBRA: devil's offer to the mathematician says

Sir Michael Atiyah: 188-190; vs. Geometry: 187-190.

ALGEBRAISTS: as with myself view mechanics & Physics as fundamentally algebraical: 188; per Sir Michael Atiyah, they follow Leibniz: 187; vs. geometers: 187-190..

ALICE'S WONDERLAND: every reality is outside fantasies such as that & continuous time: 100; every reality is outside fantasies such as that & physically extended space: 100-101.

ALPHATOPON: 3, 12 ftn. #1, 31, 35, 36, 55; calculation of: 31; correlated with acts of consciousness: 57.

ALPHAKRONON: 12 ftn. #1, 24, 31, 33, 35, 37, 45, 46, 55, 65 ftn. #1, 97, 160, 193; as objectively relative universal time: 97, 193-194; calculation of: 31; correlated with acts of consciousness: 57; Esoptrics' basic unit of time because fastest rate of change from primary act to primary act = 1 change per 1 alphakronon: 31-32, 88; with regard to astronaut paradox: 193-194.

ALZOG, REV. JOHANN: on Hugh of St. Victor & contemplation: 91.

ANGELS: do in a sense move the planets: 46 ftn. #1, 159-160; forms without carrying generators: 20, 106, 106 ftn. #1; nine choirs of: 83.

ANTI-GRAVITY: brief mention: 11; detailed: 46-47.

ANTITHESIS: also see: SYNTHESIS; also see: THESIS; science as such relative to Dialectical Philosophy: 146.

AQUINAS: see THOMAS AQUINAS, ST.

ARBITRIA: blind sculptor as example of how they work: 123-124; can't be observed without aid of sense images: 125; cause of intentionality: 124 ftn. #1; objective vs. subjective ones: 125-126; reflexively encountered acts of will whereby the encounterer focuses his power of attention: 123-125; the means by which the encounterer expresses his encountereds: 124; the "stuff" of which concepts & abstract knowledge are made: 124..

ARISTARCHUS OF SAMOS: early supporter of Heliocentrism: 1 ftn. #2.

ARISTOTLE: also see: ARISTOTLE ON INDIVISIBLES; accepted Geocentrism: xiii; disliked Plato's dialectical method: 198; for all his call for Science instead of Dialectics, he wound up using far more of the latter than the former: 198; his mentality acuity many times greater than mine: 211; his vs. Plato's use of "motion": 130 ftn. #1; like everyone else prior to Esoptrics, knew only one kind of point: 204; on "in succession": 207-208; mentioned by Hawking: 76; on Pythagoreans: 1 ftn. 2; on The Prime Mover: 113; Patristic Christians underestimated his importance says

Turner: 146; quote on anything infinite in universe: 13-14, 96; recognized unreliability of sense images: 198; rejected notion of space as nothingness: 191; says Earth is spherical: xiii, 1 ftn. #2; talks like a scientist in describing scientific vs. dialectical method: 98, 146.

ARISTOTLE ON INDIVISIBLES: also see: ARISTOTLE; his argument in his words: 204-205; his argument stands only as long as one is unaware of Esoptrics' 2 nd kind of point: 209, 210-211; his is not an argument against stop & go locomotion: 210; his lengthy argument leaves out half the moments involved: 205; master the *two* kinds of points & it's obvious he's left out the 4 durationless moments of time: 205; refutation #1 of his argument: 208-209; refutation #2 of his argument: 210-211; what he first calls moments of locomotion he later calls moments of rest: 205; what he names moments of rest are moments in which primary acts are occurring: 205; rephrased version of his argument: 210; ultimately proves only that what can be explained solely in the terms of *two* concurrent streams of *two* kinds of indivisibles cannot be explained in the terms of *one*: 211; with no knowledge of the 6 modes of power, their 2^{256} intensities, & their ways of relating to one another, he could not have had the slightest inkling of how there could be a point undivided & indivisible in space but divided & divisible in principle, but, after Esoptrics, no one can now fall back upon the same excuse: 211; with his & many another's genius greatly superior to mine, why did it take the world so long to uncover what I uncovered as a mentally inferior, uneducated youth: 211, 211 ftn. #1.

ARNOLD, THOMAS: according to Sir Michael Atiyah, he, as follower of Newton, held Physics to be geometrical: 188.

ASTRONAUT, PARADOX OF: 194-195.

ATIYAH, SIR MICHAEL: "Faustian" offer: 188; Geometry is essentially static: 189; lecture on math in 20 th century: 185, 185 ftn. #3; links Quantum Theory with "infinite-dimensional space": 186; lists theories having a long way to go yet: 190; no Algebra in a static universe: 189; on Algebra vs. Geometry: 187-190; orthogonal aspects: 187, 187 ftn. #1; references to Einstein: 185, 186; refers to general & not special Relativity: 186; refers to Thomas Arnold's claim Physics is geometrical: 188; rejected so vehemently, he had to wear a bulletproof vest: 190; says algebraists often use Geometry's pictures & I obviously do: 187-188; says geometers follow Newton & algebraists Leibniz: 187; says Geometry is concerned with space but Algebra with time & se-

quences: 187-188, 189; seems to say Relativity vs. Quantum Theory = 4 dimensional time-space vs. infinite-dimensional math: 187; uses "quantum" 23 times but never connects it with Relativity: 186.

ATOMIC CLOCKS: do not allow one to observe that time is continuous: 90, 90 ftn. #2.

ATOMS: how Esoptrics allows for 127 kinds: 47; number in Universe: 43 ftn. #1.

ATTENTION, POWER OF: see: POWER OF ATTENTION.

AUGUSTINE, ST.: my version of his "If I'm in error I am": 128; part of the Patristic era's preoccupation with mysticism says Turner: 146.

AVILA, ST. TERESA OF: See: TERESA OF AVILA, ST.

B.

BALLOON: deflated vs. inflated: 43-44.

BASIC REALITIES: also see: CARRYING GENERATORS; also see: PIGGYBACK FORMS; also see: ULTIMATES; among them, mirrored images are not equally real: 132, 133-134; are necessarily triousious in structure: 128, 138; Leibniz's monads vs. Esoptrics' hebdomads & quartuordecimads: 103, 106; outside of time vs. outside continuous time: 79, 100; potentiality of is not a mere abstraction: 139; potentiality vs. actuality of: 138; Prof. D equates mine with those of Leibniz: 79; vs. ultimate constituents: 103, 106.

BEING: meaningless term these days for most: 3, 85, potential, actual & accidental: 140; think of it as an unusual kind of energy or power: 85.

BENCHMARKS, HEAVENLY: see: HEAVENLY BENCHMRKS.

BERGSON, HENRI: unlike him I don't say change is the only reality: 130 ftn. #1.

BERNARD, ST.: 91.

BIENER: author of Internet article arguing Newton changed his view of what space is: 191-192.

BILLIARD BALL: basically most people's way of thinking of the universe's ultimate constituents: 130 ftn. #1, 195.

BITTLE, CELESTINE O. P.: God's essence = ground of intrinsic possibility: 135.

BLACK HOLES: boxes floating inside one another: 11-13; cosmic super matrices: 38; many forms made concentric by acceleration: 11, 197; mini: 11; only Esoptrics explains them: 20 ftn. #1.

BLESSED TRINITY: Esoptrics is not an attempt to deduce the mystery by human effort: 135 ftn. #1.

BLIND: from conception can't imagine sights: 81.

BLIND SCULPTOR: as example of how the arbitria work: 123-124.

BLUNDERS, COSMOLOGICAL: xiii, xv.

BODIES: most primitive sense of: 17; parts of sensible ones are logically outside of one another in the order of sequence rather than physically outside of one another in space: xiv, 16, 30 62; sensible vs. ultimate: 17.

BOHR, NIELS: as quoted by Polkinghorne: 74.

BOXES: floating inside one another as example of forms made concentric by acceleration: 11-13.

BRAIN, HUMAN: size of: 56, 57, 69.

C.

CALCULUS: I have no knowledge of it & perhaps not enough brain power to be able to learn any of it: 193.

CAMBRIDGE DICTIONARY OF PHILOSOPHY: on Rene Descartes and *res extensa*: 101 ftn. #1; on Leibniz & the continuum: 101 ftn. #2; on Leibniz & the monads: 103.

CARRYING GENERATORS: also see ULTIMATES; also see POWERS, SIX; basic description: 3, 30-31, 62, 65-66, 80; being in potency to a form gives each 6 modes of power: 4, 21, 107; being in potency to a form limits each to what primary acts it can perform: 66; exceed light speed inside atom: 44-45, 90 ftn. #1; "depicted": 5-9, 22-23, 32-33, 44-45, 63, 105, 106; generate units of tension: 30, 66, 109; how affected by pulsing of forms: 45; in potency to a form whose current OD is greater than that of their piggyback form: 9, 107; intermittently outside of time as we experience time: 80-82; in their primary acts actualize 1 to 3 of the 6 modes of power: 4, 20, 115, 196; locomotion of, is rectilinear only: 53; none moves itself to change from act to act: 113; size of: 86 ftn. #2; never move diagonally: 111-112; the 2 ways they relate to forms: 4, 21, 66; their centers always at the center of the form to which they are in potency & not at the center of their primary acts: 9, 105; time for them speeds up as their velocity increases: 37, 87.

CARTAN, ELIE: mentioned by Atiyah: 185.

CATEGORIES, REVERSE: 38-43.

CERTITUDE: defined: 122 ftn. #1.

CLOCKS, ATOMIC: see ATOMIC CLOCKS.

COLERIDGE, SAMUEL: all revolutions coincide with rise & fall of metaphysical systems: 78.

COLLISION: among generators: 37.

COLUMBUS, CHRISTOPHER: xiii.

CONSCIOUSNESS: also see: OBSERVATION; 56-59; act of a form circa OD2^{155}: 56, 68, 194; acts of, correlated with alphatopons & alphakronons: 56-57; acts of, correlated with light speed: 57-58; as a strobe light: 59, 69; deceitful God's

crippling deception: 91-92; how aware of change: 59; inventory of the content of its acts: 119-128; no awareness of duration if no awareness of change: 24-25, 58-59, 69, 91; number of acts per second: 56, 68; purpose behind no awareness of duration: 92-94.

CONTINUOUS: sometimes used to signify what's stop & go: 209-210; to be truly such every time span, locomotion, area of space, & spatial object must contain an infinitely small segment which as such is absolutely indivisible: 211 ftn. #2.

CONSTITUENTS, ULTIMATE: see ULTIMATES.

CONTACTABLES: all technological unencounterables other than the encounterer: 122-123.

CONTEMPLATION: also see: PHILOSOPHY, MYSTICAL; better tool than microscopes & telescopes: 60; key to perfect Science says Hugh of St. Victor: 91.

CONTINUUM: also see: LEIBNIZ; 15; & the prime abstractable: 126; a labyrinth according to Leibniz: 101 ftn. #2.

COPERNICUS, NICOLAUS: 1; didn't prove geocentric theory wrong but sounded its death knell anyhow: 27; Esoptrics hopes to do the same for physical extension & infinite divisibility: 28; example of moving experienced motion closer to home: 126.

CORROBORATION: inferences need it: 2; infinite divisibility needs it: 14 ftn. #2, 15.

COSMIC HISTORY: see HISTORY, COSMIC.

CO-TENANT: also see: OBJECT, POLYOUSIOUS; one of the subjects inextricably included within the confines of the same one whole: 129.

CROC: of the well-known stuff: 72, 76.

CROSS: vs. Star of David: 143 ftn. #1.

CUBES: also see PIGGYBACK FORMS; also see CARRYING GENERATORS; drawn as squares because I lack the skill to draw cubes: 5, 21; figurative way to describe forms & generators: 3, 4, 21; forms as permeable & generators as impermeable ones: 30, 31, 33, 34; primary vs. secondary: 30, 33, 34, 36, 40, 46, 63; primary = primary act generators can perform: 65, 108; subordinate noncomposite ones vs. master composite ones: 197.

CUBES VS. SQUARES: star of David vs. cross: 143 ftn. #1.

CURVILINEAR MOTION: see: MOTION, CURVILINEAR.

D.

DARK MATTER: brief mention: 68; Esoptrics on 2 kinds: 47-53.

DASHES & DOTS: 83, 84; ratio of dots to dashes

varies of the rate of change from primary act to primary act: 84.

DAVID & GOLIATH STORY: 175.

DEATH: effect on soul's OD: 93 ftn. #2.

DEATH KNELL: 28.

DEMSKE, JAMES: many say Metaphysics can't provide an objective grasp of the way things really are: 224.

DERVISH, WHIRLING: allegorical way of describing primary acts: 130, ftn. #1.

DESCARTES, RENE: my version of his "I think therefore I am": 128; *res extensa*: 101; viewed as mistaken by Leibniz: 102; without God's incessant influence there's no *res extensa* in Esoptrics: 104-105.

DEUS EX MACHINA: geometers view of space: 189.

DEVIL: offers mathematician Algebra in trade for his soul says Sir Michael Atiyah: 188; offers mathematician Geometry in trade for his soul says Haas: 189.

DIALECTICAL PHILOSOPHY: see: PHILOSOPHY, DIALECTICAL.

DIAMETER: a term Esoptrics uses only for convenience: 26 ftn. #2.

DIONS: alephons native to either OD2 or OD4: 155; as a super matrix each is 3/3 non-neutral & is a sub-sexton: 155-156; in contrast to sextons each has a single leading form native either to OD2 or OD4: 172-173.

DIRECT ENCOUNTEREDS: see: ENCOUNTEREDS, DIRECT.

DIVINE ROTATIONS: see: ROTATIONS DIVINE.

DIVISIBILITY: also see: INDIVISIBILITY, ABSOLUTE; also see: INFINITE DIVISIBILITY: also see: SEPARATION; finite only: xiv; in principle, means that in which two or more principles are conjoined: 202-203; in space vs. in principle: 201; two kinds illustrated in the contrast between Aristotle's moments D, E, Z & Esoptrics' moments C, F, H & J: 206-209; what's not *actually* divisible in space in fact may be *figuratively* divisible in space with diagrams: 203.

DRAG INDUCING DIFFERENTIAL: 17 ftn. #1, 66, 139, 140.

DRAWINGS OF ULTIMATES: not pictures of what they look like but merely a convenient way to symbolically express & communicate complex abstract principles: 3, 4-5, 103-104, 136-137, 143, 147, 156.

DULLES, AVERY: many say Metaphysics can't provide an objective grasp of the way things really are: 224.

DURATION: of acts of consciousness cannot be experienced: 24-25, 58-59, 69, 91; of changeless act can be *relatively* measured by time: 88; of changeless act cannot be *absolutely* measured by time: 24, 31; of this the universe's ninth epoch: 89; timeless moments of, vs. durationless moments of time: 83.

E.

EARTH: as flat xiii; as motionless center of universe: xiii, 1; at its center is a cosmic super matrix: 157; duration of its orbit as explained by Esoptrics: 160; example of extra-mental object divisible in space in thought but not in fact: 201; how its orbit is affected by the pulsing of the form governing that orbit: 165-166; how its orbit is stop & go without appearing jerky: 160-161; leading form of, is in potency to a form concentric with Sun's leading form: 158, 158 ftn. 1; native & current OD of the form ruling Earth's orbit: 158; rotations around Sun in primary & secondary planes: 161-164; sense in which hell is at its center: 94 ftn. #1; why it rotates around its axis: 158.

EGOMANIAC: I'm 100% guilty: 80, 99.

EIGHTFOLD WAY: 43.

EIGHTH GRADE: friend recently insisted these are the same thoughts I was expressing even in those days: 211 ftn. #1.

EINSTEIN, ALBERT: embankment & the railway carriage: 127, 127 ftn. #1; EPR Paradox: 62; Esoptrics agrees with, on time varies with velocity: 37, 87, 193; I-time: 97; mentioned by Atiyah: 185; on Universe's diameter as 10^{30} cm.: 34 ftn. #1; origin of matter: 12; quoted regarding fundamental ideas vs. formulas: 28; space is modified by mass & time by velocity: 192; spooky action at a distance: 62 ftn. #1, 76 ftn. #1, 152; supplanted Newton after improved telescopes showed Mercury's orbit did not obey Newton's principles: 198; vs. Esoptrics on fourth dimension: 92-93, 186; vs. Esoptrics on time-space near matter: 12; vs. Newton on absolute time & space: 191; vs. Newton on gravity: 12; what space truly meant for him is still a mystery for me: 192.

ELECTRO-MAGETIC RADIATION: 12.

ELECTRON: alternately moves toward & away from the nucleus: 181; concentric forms native to OD1: 11, 38, 48 ftn. #1, 93 ftn. #1; I can't decide whether all are alephons or all anti-alephons: 152 ftn. #2, 156; Science's vs. Esoptrics' explanation of why some electrons are further from atom's center than others: 181; shells: 179-181; shells vs. Esoptrics regions = chart #17: 180.

ELEPHANTS: arbitria vs. intentionality: 124 ftn. #1

EMPIRICAL EVIDENCES: see: EVIDENCES,

TIONS; must make predictions: xiv, 74, 75; not always so: 75-76, 76 ftn. #1.

THESIS: also see: ANTITHESIS; also see: SYN-THESIS; Dialectical Philosophy as such: 146.

THOMAS AQUINAS, ST.: act-oriented thinking: 15; eternity vs. time: 77; many times smarter than I am: 211; often carnal minded: 55; only in God are existence & essence identical: 81, 132, 134; on The Prime Mover: 113; Plato's vs. Aristotle's use of "motion": 130 ftn. #1, tenth crystalline sphere: 89 ftn. #1; though more a dialectical than a mystical philosopher, his help was critical to this mystical philosopher: 146.

THRESHOLD, MIRROR: see MIRROR THRE-SHOLD.

TIME: also see: TIME SPAN; absolute: 65, 65 ftn. #1; a case where all admit it's a durationless instant: 209; as measure of change: 77, 81, 82 ftn. #1; at OD1 vs. at OD2^{155}: 58, 69; by-product of change & not vice versa: 77; cannot *absolutely* measure the duration of a primary act: 24-25, 31-32, 77, 88-89, 100, 193-194; can *relatively* measure the duration of a primary act: 88, 89; doesn't apply to ultimates as long as they're performing the same one primary act: 24, 114-115, 127-128; formula for calculating objectively relative time: 194; how affected by acceleration: 37, 87; infinite divisibility of erroneously taken as empirical fact confirmed by observation: xiii, 24; issue is *what kind* of time is real & not whether or not time is real: 94 ftn. #1; I-time vs. it-time: 97; measures rate at which ultimates change from primary act to primary act: 74; objective vs. subjective: 97-98, 192-194; pulses at 2^{385} rates: 74; Science can't say what time *actually is* but Esoptrics can: 74; smallest segment for Esoptrics: 24, 115 ftn. #1; time-space strata: 9; two kinds briefly mentioned: xiv; variable vs. invariable kind: 193-195; vs. history: 94-95, 114-115; vs. time span: 209-211; what measures change from past to future can't also measure what makes no such change: 100; which measures the duration of a changeless act is a total mystery to us: 77, 82 ftn. #1.

TIME, OBJECTIVELY RELATIVE: formula for calculating: 194; vs. subjectively relative: 97-98, 192-195.

TIME SPAN: an abstraction in which 3 principles are conjoined: 204; for Aristotle is DEZ when it should be CDFE.HZJ: 208; vs. time: 209-211.

TIME VS. CONTINUOUS TIME: every reality is outside the latter for the same reason they're outside of Alice's Wonderland: 100; I never said the former is not real & ever said only the latter is unreal & merely phenomenal: 79, 94, 100; outside the latter vs. the former: 79, 100.

TRANSUBSTANTIATION: 60 ftn. #1.

TRINITY, BLESSED: see: BLESSED TRINITY.

TRIOUSIOUS OBJECT: see: OBJECT, TRI-OUSIOUS.

TRIPLICATION IN NATURE: Science wonders about it but Esoptrics does not: 176.

TURNER, WILLIAM: his article on Scholasticism in *The Catholic Encyclopedia*: 146.

TWAIN: shall never meet: 50, 156.

U.

ULTIMATES: also see: ALEPHONS; also see: BASIC REALITIES; also see: CARRYING GENERATORS; also see HEBDOMADS; also see: PIGGYBACK FORMS; also see: QUAR-TUORDECIMADS; any 2 anywhere in a logical universe can instantly switch places: 65; as bodies vs. as primary acts: 15; can meet head-on without colliding: 67-68; drawings are not pictures of what they look like: 3, 4-5; 103-104; intermittently outside of time as we experience time: 80-81, 88, 100; microscopic vs. macroscopic: 3, 31, 62, 64, 80; no 2 simultaneously activate the same level of power: 19; not involved in time while performing the same one primary act: 24; outside of time vs. continuous time: 79, 100; "size" of is only figurative: 3; rates of change from primary act to primary act vary: 17; separate due to uniqueness of primary acts: 17; separateness illustrated by a line 0A: 18; separateness illustrated by a line B0A: 19; smallest size of: 86 ftn. #2; unimaginable vs. unknowable: 15.

ULTIMATE CONSTITUENT: one form & one carrying generator forever joined at the centers of their actuality: 3, 66; vs. basic reality: 103, 106.

UNIFIED FIELD: non-singular solution: 12;

UNIVERSE: also see: SPACE; as a multi-verse: 27; a single point: xiv, 16, 30; astronomical collection of ultimates taken as a whole: 84; collection of forms as space envelopes: 27; collection of kinds of primary acts which any generator can perform: 32, 62-63, 196; duration of according to Esoptrics: 24, 37, 89, 92 ftn. #1, 95 ftn. #1; form of: 10, 34, 38, 39, 40, 47, 89; halfway thru its history is the same for all but how much time it takes to reach that point is not the same for all: 94-95, 95 ftn. #1; major vs. minor: 93; micro- vs. macro-constituents of: 3, 31, 62, 66, 80, 81; ninth epoch of: 16, 84, 86; no change of any kind anywhere in it below 10^{-96} seconds: 115 ftn. #1; nothing in it in any way infinite: 14; number of hydrogen atoms in: 43 ftn. #1; number of galaxies in: 42 ftn. #1; path of light in a physically vs. a logically extended one: 53; physically a singularity: 16, 24,

30, 62, 65; rotation of: 89-90; tenth epoch of: 86 ftn. #1; Wikipedia on age of: 43; Wikipedia on diameter of: 42; Wikipedia on number of hydrogen atoms in: 43 ftn. #1; Wikipedia on number of galaxies in it: 42 ftn. #1; will only expand in the ninth epoch but contract in the tenth: 86, 86 ftn. 1.

UNIVERSE, DIAMETER OF: according to Esoptrics = 10^{31} cm.: 16, 42 ftn. #1; how greater than what light speed allows: 43-46; Wikipedia on: 42.

UNIVERSITIES: a thousand sent copies of document #3: 61 ftn. #1.

V.

VELOCITY: Esoptrics' formula for: 47; increase in slows time for forms but opposite for generators: 37.

W.

WATER SKIER: as example of how patterns of divine rotation produce magnetic lines of force: 167-178.

WAVES VS. PARTICLES: 68.

WEBSTER'S UNABRIDGED DICTIONARY: on the meaning of "continuous": 209.

WHIRLING DERVISH: allegorical way of describing primary acts: 130, ftn. #1.

WIKIPEDIA: age of Universe: 43; diameter of Universe: 42 ftn. #1; eight-fold way: 43; electron shells: 179-181; EPR Paradox, 62 ftn. #1; number of atoms in universe: 43 ftn. #1; number of galaxies: 42 ftn. #1; size of human brain: 56.

WITTGENSTEIN, LUDWIG: mentioned by Hawking: 76; Philosophy's only task now is analysis of language: 76; wrong about Philosophy's limits: 98.

Z.

ZENO OF ELEA: xiii, 2 ftn. #1, 14, 15, 90 ftn. #2; Achilles & the tortoise: 25; as evidence consciousness is not a *crippling* deception: 91; when first I learned he agreed with me: 96.

✡✝✡✝✡✝✡✝✡✝✡✝✡✝✡✝✡✝✡✝✡✝✝

Printed in the United States
by Baker & Taylor Publisher Services